IET CIRCUITS, DEVICES AND SYSTEMS SERIES 24

Fundamentals of Electromagnetic Levitation

Other volumes in this series:

Volume 2 **Analogue IC design: the current-mode approach** C. Toumazou, F.J. Lidgey and D.G. Haigh (Editors)
Volume 3 **Analogue-digital ASICs: circuit techniques, design tools and applications** R.S. Soin, F. Maloberti and J. France (Editors)
Volume 4 **Algorithmic and knowledge-based CAD for VLSI** G.E. Taylor and G. Russell (Editors)
Volume 5 **Switched currents: an analogue technique for digital technology** C. Toumazou, J.B.C. Hughes and N.C. Battersby (Editors)
Volume 6 **High-frequency circuit engineering** F. Nibler *et al.*
Volume 8 **Low-power high-frequency microelectronics: a unified approach** G. Machado (Editor)
Volume 9 **VLSI testing: digital and mixed analogue/digital techniques** S.L. Hurst
Volume 10 **Distributed feedback semiconductor lasers** J.E. Carroll, J.E.A. Whiteaway and R.G.S. Plumb
Volume 11 **Selected topics in advanced solid state and fibre optic sensors** S.M. Vaezi-Nejad (Editor)
Volume 12 **Strained silicon heterostructures: materials and devices** C.K. Maiti, N.B. Chakrabarti and S.K. Ray
Volume 13 **RFIC and MMIC design and technology** I.D. Robertson and S. Lucyzyn (Editors)
Volume 14 **Design of high frequency integrated analogue filters** Y. Sun (Editor)
Volume 15 **Foundations of digital signal processing: theory, algorithms and hardware design** P. Gaydecki
Volume 16 **Wireless communications circuits and systems** Y. Sun (Editor)
Volume 17 **The switching function: analysis of power electronic circuits** C. Marouchos
Volume 18 **System on chip: next generation electronics** B. Al-Hashimi (Editor)
Volume 19 **Test and diagnosis of analogue, mixed-signal and RF integrated circuits: the system on chip approach** Y. Sun (Editor)
Volume 20 **Low power and low voltage circuit design with the FGMOS transistor** E. Rodriguez-Villegas
Volume 21 **Technology computer aided design for Si, SiGe and GaAs integrated circuits** C.K. Maiti and G.A. Armstrong
Volume 22 **Nanotechnologies** M. Wautelet *et al.*
Volume 23 **Understandable electric circuits** M. Wang

Fundamentals of Electromagnetic Levitation

Engineering sustainability through efficiency

A.J. Sangster

The Institution of Engineering and Technology

Published by The Institution of Engineering and Technology, London, United Kingdom

The Institution of Engineering and Technology is registered as a Charity in England & Wales (no. 211014) and Scotland (no. SC038698).

First published 2012

The Institution of Engineering and Technology
Michael Faraday House
Six Hills Way, Stevenage
Herts, SG1 2AY, United Kingdom

www.theiet.org

British Library Cataloguing in Publication Data
A catalogue record for this product is available from the British Library

ISBN 978-1-84919-663-5 (hardback)
ISBN 978-1-84919-664-2 (PDF)

Typeset in India by MPS Limited
Printed in the UK by CPI Group (UK) Ltd, Croydon, CR0 4YY

To my grandchildren

With the growing unpredictability of life in the 21st century,
as a result of climate change, the future security
of your generation will surely derive from
the resilience gifted by education.

Contents

Preface

The growing, corroborative and consistent evidence, emanating from the peer-reviewed literature on climate science, is that mankind is ignorantly initiating a climate change tsunami by wilfully continuing to combust fossil fuels despite half a century of warnings. As Joel Achenbach has trenchantly observed:

> In the same way that the 20th century was the century of world wars, genocide, and grinding ideological conflict, the 21st will be the century of natural disasters and technological crises and unholy combinations of the two. It'll be the century when the things that we count on to go right will, for whatever reason, go wrong.

What we will not be able to depend upon is the continuation of consistent weather patterns that have been familiar to our forefathers over many generations. Food production will become extremely difficult, and much of the farmed produce that we have become used to will no longer be available. It is difficult to see how the teeming billions of humans, projected to rise beyond 10 billion by 2100, will feed themselves once climate change truly takes effect. At worst, if a runaway greenhouse process were to take hold of the climate system then a hot arid and dead Venus-like planet is the medium-term future for the earth. As James Hansen, a climate scientist at NASA, has observed:

> I've come to conclude that if we burn all reserves of oil, gas, and coal, there is a substantial chance we will initiate the runaway greenhouse. If we also burn the tar sands and the tar shale, I believe the Venus syndrome is a dead certainty.

However, this dire future can still be averted, while maintaining a relatively advanced and civilised planet, by making a very rapid and early transition to world economies powered by renewables. Many recently published books have proffered route maps to a sustainable future based on the rapid adoption of renewable energy, and many expect that a concerted and determined drive, to improve efficiency in the technology sectors of advanced economies, must form a major component of any strategy aimed at procuring a successful transition away from fossil fuels.

Once the profligate and wasteful uses of energy have been eliminated from human activities, there will be a continuing need to strive for efficiency as we come to terms with life on an energy-limited planet dependent on renewables. This will have a profound impact on engineering design and development. Commercial aircraft may no longer disturb the troposphere but people will still fly, at very high speed, three or four centimetres above a track of electromagnets, possibly in evacuated tubes to reduce air resistance. The energy required per traveller to move at three or four times the speed of sound in a frictionless tube will be much lower than lifting them to 40,000 feet to travel at the speed of sound in what is essentially a projectile. The route to this future is through electromagnetic levitation or suspension. For a long time this technology was little more than an aid to fair ground illusionists and magicians. However, the technology has advanced significantly since those days. For example, magnetically levitated and active bearings have now permeated into a wide range of technologies from flywheels to blood-pumps, from photolithography to medicine and haptic systems. The same active technology is also used in systems that employ linear electric drives. These are now most commonly and popularly recognised in the form of 'maglev' trains.

This text anticipates these impending engineering imperatives to improve efficiency of electrical machines and devices. The desire to minimise frictional and bearing losses and ohmic losses in conductors, the major causes of inefficiency in machines, will be a paramount goal. Solutions are likely to revolve around levitated friction-free motion, thus creating oil-free bearings and wheel-free rail systems while promoting the use of superconductors wherever possible.

The book seeks to provide coherent and wide-ranging instructional material on electromagnetic levitation and suspension by collating all of the currently available developments in this technology sector embellished with enough mathematical detail and discussion to enable the reader to fully comprehend the basic physics. As far as the author is aware, the full range of levitation technologies has not previously been presented in a single textbook. Most alternative texts, if they are still in print, concentrate on the impact of levitation techniques on transport. It is perhaps pertinent to note here that the levitation/suspension of an untethered object raises difficult sensing and control issues, particularly for obstacles freely suspended in an unconfined space (six degrees of freedom). Fortunately, within practical machines, the moving parts, which require to be made friction-free, are seldom totally unconfined, and consequently the control problem is, more often than not, one- or two-dimensional in character, rather than six-dimensional. So, where stability is addressed, within the chapters of this book, it is largely in this restricted form.

Ten chapters are required to deliver all of the significant aspects of the still-developing electromagnetic levitation story. There are eight chapters focused exclusively on identifiably separate aspects of the topic. These are sandwiched between an introductory chapter, which presents the reader with the plentiful evidence of the profligate and harmful consumption of cheap fossil-fuel energy by advanced economies, and a concluding chapter advocating an escape route from fossil fuels, through efficiency, renewable energy and friction-free technology, all aimed at repairing the biosphere before it is too late. Some of the observations on efficiency in Chapter 1 have been extracted from the author's book entitled *Energy for a Warming World*, with kind permission from Springer Science+Business Media B.V. The vast majority of the figures and diagrams are original to the author. The few that are not have been extracted from the published literature in which case the sources are fully referenced.

Chapter 1
Introduction – thirst for energy

Engineering is the practice of safe and economic application of the scientific laws governing the forces and materials of nature by means of organization, design and construction, for the general benefit of mankind.

S.E. Lindsay (1920)

1.1 Fossil-fuel era

1.1.1 An age of engineering inefficiency

The fossil-fuel era is coming to an end. Thus far, it has spanned about 250 years from 1760 to 2011, and according to the historical record, these years have represented a period of immense change in technological terms, encompassing brilliant scientific breakthroughs and engineering innovation. But, what history generally fails to report is that it has also been a period of inexcusable profligacy in relation to our treatment of the planet's initially bountiful energy supplies. S.E. Lindsay's desire (see above quotation), in 1920, for the responsible delivery of engineering progress has not occurred. This is largely due to the fact that, despite their finiteness, the primary sources of power – coal, oil and gas – have been inordinately cheap throughout this time span, since these products of ancient sunlight, and the pollution they cause when combusted, have largely been treated as uncosted 'externalities' in the classical economic science, which underpins our global capitalist system.

When viewed from a strictly electrical engineering perspective, which is of particular interest here given the thrust of this book, developments in electrical devices on the one hand – such as white goods, power tools and computers – and systems on the other – such as the electricity grid – have generally been designed and implemented with only passing interest in the need for efficiency. Where it occurs, it is usually driven by the desire to procure competitive advantage, rather than by the minimisation of losses. The simple reason for this is that power station fuel has been plentiful and ridiculously inexpensive, resulting in artificially low prices for electricity. If there is an exception to this trend, it lies in the modern high-speed communications sector where device efficiency and low losses are critical to maximising signal transmission and reception coverage zones, and to maintaining temperature levels within safe operating ranges in high-density integrated circuits. We will draw inspiration from the developments in this sector later in the book.

The early history of electrical engineering is well documented in many textbooks [1,2], and abundant information on past developments can be found on the Internet. Following early pioneering work, by an extensive list of contributors (see Chapter 2), Michael Faraday (1791–1867) was the first to demonstrate a practical electric motor, and he did this in 1821. As we shall see, the developmental history of the motor/generator provides a perfect example of the lack of attention to electrical efficiency by early design engineers.

There followed a flurry of electrical innovation and development over the next 40–50 years, culminating in a direct current (DC) electric power generation system to supply mercury arc lighting in San Francisco. This appeared in about 1880. At about the same time more versatile alternating current (AC) electrical supply systems were being built in Europe, such as the 100 mile long, 30 kV AC transmission line that was erected in Germany in 1891. These early generators were either powered by water turbines in hydro-systems, or, more commonly, by coal-fired steam turbines. It is clear from the history that engineers were exercised by boosting power capacity, increasing flexibility and establishing reliability, with efficiency being very much an afterthought.

1.1.2 Profligate energy usage

Of course, the growing availability of electricity spawned a whole new category of consumers of power and continues to do so even today. In the early days, this category included suppliers of lighting and heating systems, users of compact and versatile machine tools, and operators of electrical telegraphy and telephony. Availability of electricity was made possible by copiously burning cheap coal, consumption of which continued to rise exponentially. In the transport sector, and in America, the first electrification of a steam railroad occurred in 1895, but it is doubtful if these early electric trains were any more efficient in their use of coal (at the electrical power station) than the steam locomotives that they replaced, but this was hardly an issue since efficiency was not given much importance. Coal power stations at the time were little more than 20–25% efficient, while steam turbines, electrical generators, transformers, electricity transmission lines and the locomotives themselves all exhibited much higher resistive heating losses than could have been achieved with careful design. Consequently, at the point of traction, between the locomotive's wheels and the rails, the efficiency of the use of the chemical energy embedded in the molecular structure of the original coal was likely to have been little more than 5%. So 95% of the energy in the incinerated coal was uselessly polluting and warming the local environment. In truth this was little different, in environmental terms, from burning all of the originally mined coal on open fires! Little wonder 'global warming' is now a problem. But coal was so inexpensive that this gross level of inefficiency was not considered to be a hindrance to continued development.

The above saga of wasted power continues to be replicated in today's electrical supply systems. Losses in the transmission of electrical power across the grid are estimated to be of the order of 7% in the United States and Europe. Consequently, at the point where electricity is distributed to end users, only between 30% and 50% of the original power supplied to the system, in the form of heat from steam in a conventional generating station, is available. The two figures depend on whether an estimate of 36% or 60% is adopted for turbine efficiency. This implies that at best only *half* of the energy in the coal or oil burned in a modern fossil-fuel-fired power station produces useful electricity for the consumers. The rest almost literally goes 'up in smoke'!

1.1.3 End-user wastage

The situation is actually worse than this if we consider end-user applications, which are rarely efficient. Returning to the electrical power leakage problem, we can confidently state that the primary end users of electricity are industry (21%), energy supply (6%), commerce (13%), public sector (35%) and homes (25%). In industry we also have the statistic [3] that almost 66% of the electricity is used to drive motors and supply transformers. The efficiency of electrical devices of this genre [4] is of the order of 90%. Most of the remaining 34% of electricity distributed to industry can reasonably be assumed to be directed towards lighting. Again presuming that industry largely uses fluorescent lighting, we can estimate that their lighting efficiency (proportion of input power converted to light) is probably of the order of 10%. So we can say that industry wastes $0.1 \times 0.66 + 0.9 \times 0.34 = 0.372$, that is 37% of the electrical power supplied to it. Of course the lighting problem, and the high power loss that causes it, is being slowly resolved, but it is still far from acceptable.

In commercial offices, and in the public sector (schools, hospitals, offices etc.), the vast majority of the supplied electrical power is used for lighting and air-conditioning/heating, with a small proportion used to operate computers and appliances. An estimated overall efficiency of the order of 25% is not too remote from reality, and, consequently, these sectors jettison 75% of the power supplied to them into the environment as heat. The way in which electricity is consumed in the domestic sector in the United Kingdom is available in government literature [3]. On average, post 2000, 25% is used in lighting – probably mainly incandescent bulbs. These are very inefficient with just 2% of the power used appearing as light. Twenty-four per cent of domestic electricity is employed in supplying fridge/freezers that are rather inefficient because of poor insulation, although the pump motor and power supply will also make a substantial contribution to a typically quoted figure of 12%. Washing machines, dryers, dishwashers account for 18% of usage. Here losses can be attributed to power supply transformers, pump motors, bearings and expelled hot water, giving a typical efficiency of about 60%. Cookers, kettles, microwaves take up another 18%, and while microwave ovens are reasonably efficient at 70%, cookers and kettles are not – so an overall figure here of 50% is probably not too wide of the mark. Finally, consumer electronics that have penetrated the domestic market in a big way in recent years now accounts for 15% of electrical power usage in the homes. In these electronic devices, electrical losses are mainly attributable to power supply units that can deliver efficiency levels as low as 20% and as high as 80%. A mean of 50% is probably not unrepresentative. Of the electric power supplied to homes in the United Kingdom, the power wasted is as follows:

- Lighting $= 0.25 \times 0.98 = 24.5\%$
- Fridge/freezers $= 0.24 \times 0.88 = 21.1\%$

- Washing machines = 0.18 × 0.4 = 7.2%
- Cookers = 0.18 × 0.5 = 9%
- Electronics = 0.15 × 0.5 = 7.5%

This gives an accumulated electrical power loss of at least 70% in a typical home in the United Kingdom (the number will probably apply in most industrialised nations). Space heating/cooling is ignored for this exercise since few homes (except perhaps in nuclear France) today are conditioned using electricity at the present time. The total wastage in all sectors is therefore as follows:

- Industry = (0.21 + 0.06) × 0.37 = 10%
- Commerce = 0.13 × 0.75 = 9.7%
- Public sector = 0.35 × 0.75 = 26%
- Homes = 0.25 × 0.7 = 17.5%

This represents a total percentage loss of 63% by end users. Of the power contained in the heated steam entering the turbines of a combined cycle electricity generator only, at best, 19% [(0.5 × 0.37) × 100] results in usefully employed power by the end user. That is, less than one-fifth of the power in the original fossil fuels is employed usefully! For generators operating in non-combined cycle mode the figure is even lower at 11%, with 89% of the original power being wasted.

Clearly the accumulated power dissipation involved in generating electricity, in transporting and distributing it, and in the way we use it, is almost ‘mind boggling’ in its magnitude and could only ever have arisen in an era of ridiculously cheap fuel. At present, the ‘saving grace’ is that electricity represents only 10% of the energy usage in industrialised societies. In the future when electrical power provides 100% of our needs, this level of misuse of precious resources is likely to be viewed as quite irresponsible and unacceptable! It seems utterly perverse that governments are tentatively preparing to move to a world based on renewables, supplying power to consumers through the agency of electricity, without seriously addressing the vast wastage, which is currently prevalent in the electricity industry? Many of the inefficiencies are insurmountable, of course, but obviously the more efficient we can make electricity supply and distribution, and the more efficiently we use it, the less pressure will be placed on the drive to exploit renewable resources, and the less will be the environmental damage. In this book, we shall endeavour to offer possible routes to significant improvements in the efficiency of a limited but important category of electrical equipment.

Finally, it is perhaps pertinent to note that the higher than necessary levels of atmospheric pollution, which was the inevitable consequence of lax engineering practices during the fossil-fuel era, did not trigger any discernible levels of concern, except perhaps in cities in relation to smog and health issues. Needless to say it had little impact on engineering design methods that placed a low priority on efficiency. In fact, pollution was of no consequence to most people for most of the industrial age. It only became a global issue in the early twenty-first century, when the public in general began to express concern about ’global warming’ and climate change.

1.2 Energy-constrained era of renewables

If a decisive shift to renewable sources of power for all civilisation's energy demands is implemented, hopefully by 2025, to avert the threat of seriously damaging climate change, then the need for high efficiency in the technological sphere will become imperative of engineering design and practice, because there will be a decidedly limited supply of electrical power from such sources. Cautious estimates [4] suggest that by 2050, 14 Terawatts (TW) will be extractable globally from renewables. This is not enough to satisfy business-as-usual (BAU) growth but is more than enough to operate modern global economies at the level prevalent in 1990 when consumption was about 12 TW, provided major drives to root out energy inefficiencies in all sectors of our modern economies are implemented. It seems likely that energy will become very expensive and that a probable result will be a withering of globalisation as currently practiced [4].

So how much power can renewables really provide to a rapidly growing population of increasingly energy-hungry human beings, who are predicted to consume almost 30 TW of power [5] by the latter half of this century? It is possible to show, by first assessing the magnitude of the resource as predicted by fundamental physics, followed by thermal efficiency calculations on the collected power as it is subsequently processed through various stages of electricity production – turbines, generators, up-conversion transformers, transmission over the grid, down-conversion transformers and distribution to consumers – that, despite the 'hype', the power available to users is by no means limitless [4]. 'Firm' estimates for ultimate electrical power levels, which can be extracted from *realistically* accessible renewable sources, are summarised in Table 1.1. These estimates are of an accuracy, which an engineer would describe as being of 'ball-park' reliability, since they are based mainly on engineering evaluations of the science and technology, although they are backed up by some less reliable geographical and geological guesstimates relating to the exploitable extent of the most suitable power collection sites.

Table 1.1 Global potential for renewable power at customer sockets

Renewable resource	Available power at the point of consumption (TW) (2050 +)
Hydro	~2.0
Wind	7.5
Wave	0.022
Tidal	0.2
Solar	4.5
Geothermal	0.14
Nuclear	1.8
Total	**~16.2**
Fraction of projected demand [5]	54% (~30 TW)

The observations in section 1.2.1 are perhaps apposite, given the growing endorsement of renewables as civilisation's primary source of energy.

1.2.1 Renewable energy sources

Hydro

Hydroelectric schemes represent a mature renewable energy resource, and in the Western industrialised nations most of the viable sites for reservoirs and dams have been commandeered. Growth is occurring mainly in the new industrial nations of Asia, in particular China and India. It is difficult to assess to what extent mankind will exploit sites, which in the past might have been considered to be too difficult, too controversial and too expensive, once fossil-fuel-derived energy is in very short supply. It is estimated that about 2 TW, almost three times the 2009 level, is the best that could be achieved in the long term. This is of the order of 13.4% of the power (15 TW) currently consumed by mankind.

Wind

Given the extent to which land is already being commandeered by the human population, it is estimated [4] that using currently available technology, an area of land and shore, about equal to the land area of Mexico, but spread across the globe, could possibly be identified for coverage by wind farms, if the desire for energy becomes sufficiently desperate. This results in the figure of 7.5 TW of power to the consumer from wind, after all loss mechanisms have been factored into the calculation. It is difficult to see how more could be extracted from wind in the long term if we assume that the human population will continue to grow, and will require to maintain, at least at present levels, other forms of land usage. The figure is equivalent to 50% of current (2009) demand – a very significant contribution from wind – but it is predicated upon solving the variability issues.

Wave

Wave power will contribute only a tiny fraction (0.14%) of man's energy needs even in the very long term. Despite the fact that the power in the waves is vast, little is available for exploitation, unless we learn to extract it in the deep ocean. With current technology we are limited to shore-based, or close to shore, collection schemes. In addition, global coastlines that offer good waves, in sites that are not unfeasibly hostile, are estimated to be no more than 5,000 km in extent. Even if all of this coastline were optimally employed as wave farms, the most that we can possibly harvest is about 22 GW.

Tidal

The gravitational physics governing tidal movements indicates that the potential energy built into the 'pull' of the sun and moon on the seas and oceans of the globe, while large (equivalent to about 130 Aswan dams), is quite limited by comparison with hydro, wind and solar resources. Like wave activity it is also very difficult to access. Suitable sites for barrage and tidal stream methods of tapping into the tides are scarce, and calculations suggest that, at best, 0.2 TW of electrical power could be extracted from tidal resources. This is 1.3% of current demand.

Solar

Using basic geometry and the known radiant power of the sun it is possible to establish a figure for the radiant power density striking the earth's disc. This quantity is termed the solar constant and is currently estimated to be 1,367 W/m^2. From this, the solar power density at the earth's surface can be computed and is generally quoted as having a mean value of 170 W/m^2. Used unwisely this statistic can generate hugely over-optimistic estimates of exploitable solar power levels. This is because, when delivery to the consumer is the criterion, and conversion, generation, transforming, transmission and distribution inefficiencies are factored into calculations, a figure of 4.5 W/m^2 is obtained, which can be extracted from solar radiation with currently available technology. The most effective locations for solar farms are hot, arid deserts, but even these locations have other uses and are not devoid of ecological importance. Land area available for massive solar farms is not 'unlimited' and reasoned deliberation suggests that an upper limit of 4.5 TW of electrical power is available from solar sources. In the long term, therefore, about 15% of projected global demand (30 TW [5]) could be met from solar power stations and other solar gathering activities.

Geothermal

As with wave and tidal power, geothermal power represents a useful but small resource in global terms. Reliable estimates suggest that output from this resource, with current levels of technology, could over time possibly reach 15 times the power being delivered in 2005, which gives a ball-park prediction for geothermal power of 140 GW. Consequently, geothermal sources could potentially add 0.6% of demand to the renewables 'mix' in the long term.

Nuclear

Figures for nuclear power generation have been included in table 1.1 for completeness, although nuclear fission is not strictly renewable. It is included here since it seems unlikely that it will not be needed to provide reliable base-load electricity for a global supply system particularly during the transition to a fully functioning renewable network.

1.2.2 Potential power limits

On the basis that an unforeseen technological breakthrough in extracting electrical power from renewable resources, such as nuclear fusion, is not on the horizon, and that present methods will not advance much beyond current levels of sophistication the engineering evidence strongly suggests that electrical power generated from all renewable sources, backed up by nuclear power, will, in the long term (beyond 2070), probably plateau at a 16 TW level. This equates to about 50% of a possible demand of ~30 TW, which is predicted by statistical projections to 2050 and beyond, were we to be foolish enough to continue to pursue BAU far into the century. It is presumed that human population will plateau at 10.5 billion towards the end of the century, and that mankind continues to be in thrall to an energy profligate, consumption-driven, global economic system. To improve on the 16 TW figure (Table 1.1) would either take a step-function change in technological expertise and engineering prowess, particularly with regard to operating in hostile marine environments, or an unlikely acceptance by human societies of a visual pollution level and environmental degradation thresholds, associated with covering vast areas of land and sea with wind, wave and solar farms, far beyond today's acceptable boundaries. In my long experience of engineering practice, major technological advances generally take about 20–30 years to move from an idea to full commercial implementation.

The indications are, that the coming age of renewables will be energy constrained, and therefore power will likely be expensive. In this case, it seems fair to presume that technology-based societies will expect and demand high levels of efficiency from their electrical equipment. It may seem presumptuous to attempt to envisage how, in an impending era of expensive electrical power, engineers will attempt to address the efficiency issue and how electrical machines will develop over the next 50–60 years – concepts and techniques, not yet thought of, will undoubtedly be evolved to improve efficacy, utility, versatility, manufacturability etc. However, significant improvement in the inherent efficiency of any electrical device entails addressing some basic physics relating to ohmic loss and, in a device incorporating movement, friction. Thus, these physical impediments to efficient operation, and how engineers of the future will attempt to overcome them, can certainly be assessed, well in advance of large-scale implementation.

1.3 Inefficiencies in electrical equipment

Life today is almost unimaginable without mains electricity. It provides lighting for houses, buildings, streets, supplies power for domestic and industrial heating, and for almost all electrical equipment used in homes, offices, hospitals, schools and factories. Improving access to electricity worldwide has been a key factor in 'oiling the wheels' of modern life. However, the 'story' of electricity is not all technical wizardry. It is also a story of inefficient power usage. To demonstrate this we need look no further than electricity generation and its usage as it is currently practised around the world.

In modern oil, gas, coal, hydroelectric or nuclear power stations, the generator set is usually of the synchronous type. This means, that, in the context of rotating electrical machines, the generator set is of the 'inverted' type of construction, where the windings supplying the electrical power are stationary, being wound onto the stator, while the armature (the moving part) houses a rotating magnetic stack.

(The key electrical equations governing motor and generator effect are developed in Chapter 2.) A key feature of this set-up is that the generated voltage is alternating (AC) and its frequency is strictly controlled by the rotational speed of the machine. The frequency of the supplied AC output of a given machine is not difficult to estimate. It is given by the product of the number of rotor poles (magnetic poles) and the rotational speed in revolutions per minute (rpm), divided by 120. A four-pole machine (probably the most common arrangement) rotating at 1,500 rpm will generate a 50 Hz supply voltage. In the United States where the electricity supply is set at 60 Hz, the speed of the machine has to be 20% higher. Maintaining the frequency of the AC output within acceptable limits means that the prime mover (engine or turbine) must be governed to hold its speed constant to within 3–4% of the optimum value. The generated voltage for a single-phase machine is given by a winding constant (~4.15) multiplied by the number of turns (i.e. in the stator winding) multiplied by the flux under each magnetic pole multiplied by the frequency [6,7]. A single-phase machine is one that essentially has only one stator winding with two output terminals. The supplied voltage is a single alternating signal at the design frequency of the machine. More power with higher efficiency is delivered if the stator carries more than one winding; the norm is three, in which case there are three live output terminals, plus a neutral connection, delivering three-phase power. This means that between any two terminals there is a 50 Hz sinusoidal signal as in a single-phase machine, but the phase voltages, as they are termed, although of similar magnitude are shifted in phase relative to each other by ±120°. The nature of a three-phase supply and its application is not significant in relation to the environmental impact of electrical systems, once system losses have been established, and we will not need to refer to this generation mode again, other than to provide some background for discussions on electricity transmission.

Output voltages of between 20,000 and 30,000 V are typical of synchronous machines installed in modern fossil-fuel power stations. The fundamental requirement on large power station generators to deliver about 30,000 V (30 kV) at 50 Hz (or 60 Hz) exercises a considerable constraint on the design of a synchronous machine and in visual terms structural differences between machines can seem marginal. The primary variable is power capacity, and since power is essentially volts multiplied by amperes, it means more powerful machines have to be capable of sustaining higher currents when on full load. High currents incur heat, which means bulkier windings, and more effective cooling is required to minimise energy loss. Inevitably the machines become bigger, although structural constraints of a mechanical complexion are enforcing a halt to the trend. Fossil-fuel power stations are capable of generating power levels to the grid anywhere between 100 and 2,000 megawatts (MW). If we take an intermediate figure of 500 MW, a synchronous machine capable of supplying this sort of power will be of the order of 30 ft long and 12 ft in diameter. Such a machine would typically supply 21,000 A at 24,000 V, at a frequency of 50 Hz (60 Hz) as it spins at between 3,000 and 3,600 revs/min.

All of the above sounds positive and constructive in electrical supply terms. But, unfortunately, not all of the power provided by the prime mover is being converted into electrical power to the grid within the synchronous generator. There are a number of sources of power loss that cannot be circumvented, although some effort may be made to keep them as low as possible. These unavoidable loss mechanisms are conductor losses, core losses, mechanical (frictional) losses and stray losses.

1.3.1 Electrical loss mechanisms

Conductor loss

Conductor loss, or ohmic loss, occurs whenever current is forced through a wire. It comes about because real wires formed from copper or aluminium, for example, exhibit some resistance (Ohms) to free electron flow (current in amperes) through them. Electrons flowing along a wire can very crudely be likened to balls rolling down a pin-ball machine. In the pin-ball machine the balls, as they travel, collide against the pins, which divert them from the direct path to the bottom of the table. After many collisions the total distance travelled by a typical ball will be much more than the length of the table. On colliding with a pin, a ball will actually lose a tiny amount of kinetic energy that will appear as vibrational energy in the pin. The movement of an electron through a wire is not unlike this. Fixed copper or aluminium atoms (the pins in the pin-ball machine) present obstacles to the electrons flowing through the wire. The distance travelled by the average electron is generally much longer than the direct path through the wire, and the collisions between the electrons and the fixed atoms generate atomic scale vibrations. This atomic agitation manifests itself as heat. Only at absolute zero temperature (0 K) are the atoms of a material completely still. In a generator, if the winding current is known and if the resistance of the winding has been measured, the heat loss in watts is given by amperes-squared times winding resistance [6,7] multiplied by the number of windings. For a typical synchronous generator, this copper loss is approximately 3% of the power delivered. In a 500 MW machine, 15 MW is dissipated in the windings. This is enough to boil the water in 15,000 brim full kettles, or to power, across Europe, two high-speed electric trains!

Core loss

The windings in any electrical machine are usually wrapped around cores made of soft iron. 'Soft iron' is a form of iron that is easily magnetised or demagnetised (spinning electrons are easily aligned or misaligned) and it is used to maximise magnetic flux through the windings of the machine. These cores form the armature and stator of the machine. In an operating generator the large currents flowing in the windings induce magnetic fields in the cores. These alternating magnetic fields, through Faraday affect, which is actually quite similar to generator affect, induce secondary currents within the core structure of the generator. These currents are commonly termed eddy currents. Since iron is not a perfect conductor, resistive losses associated with the eddy currents also occur within the metal (iron) structure of the generator. Most people who have ever used a battery charger would be aware that the charger gets warm. This is because the charger contains a transformer, which has an iron core encased in multi-turn windings made from insulated copper wire. The heat has the same source as in the generator – namely copper loss and core loss. If you are very perceptive, you may also have observed a faint hum coming from the charger. The laminated core of the transformer, which is laminated to minimise core loss, can vibrate (hence the hum) because of relative movement between laminations. The movement is driven by the Lorentz force described earlier (essentially motor effect). This process absorbs energy and adds to power loss. It also occurs in AC power generators and is considered to be part of the core loss of the machine. This loss is generally in the vicinity of 4% of the generator output – enough to heat 12,000 platefuls of Scots' porridge in 12,000 microwave ovens! Even if you like porridge: not a great idea.

Mechanical loss

Mechanical losses mainly include drag effects due to air compression and air friction, which occurs in the air gaps between the rotor and the stator when the rotor is revolving at high speed. Bearing losses also come into this category. In total, power dissipation of a mechanical nature can contribute a further 4–5% reduction in machine efficiency. Stray losses describe all the other miscellaneous losses that do not fall into the above 'pigeon holes', and although small they represent a finite addition to inefficiency. These losses are generally estimated to contribute about 1% to the total. A generator with a 500 MW rated output power will, because of these losses (12% of 500 MW equals 60 MW), require a turbine or diesel engine delivering 560 MW of mechanical power at its input shaft. The generator efficiency is then $(500/560) \times 100 = 89.3\%$. For most power station generators, the assumption of a figure of 90% for generator efficiency would not be far off the mark for the purposes of assessing environmental impact.

1.3.2 Thermal losses

We also need to give some attention to the 'prime mover' in any generation station. Today about 86% of the world's electricity is generated using steam turbines fuelled mainly by coal and oil. Most of the rest is provided by nuclear power that also generates electricity through the agency of steam. We can therefore limit our attention to this type of prime mover when we later use conventional power generation as a benchmark for assessing generation from renewable sources. The steam turbine is designed to extract thermal energy from pressurised steam, and convert it into useful mechanical power output. It has almost completely replaced the long-lived reciprocating piston steam engine, primarily because of its greater thermal efficiency and higher power-to-weight ratio. Also, because the turbine generates rotary motion directly, rather than requiring a linkage mechanism to convert reciprocating to rotary motion, it is particularly suited to the role of driving an electrical generator.

In thermodynamics, the thermal efficiency (η_{th}) is a dimensionless performance measure of a thermal device such as an internal combustion engine, a boiler or a furnace. The input to the device is heat, or the heat content of a fuel that is consumed. The desired output is mechanical work, or heat, or possibly both. Because the input heat normally has a real financial cost, design engineers close to the commercial realities tend to define thermal efficiency as [8] 'what you get' divided by 'what you pay for'.

When transforming thermal energy into mechanical power, the thermal efficiency of a heat engine is obviously important, in that it defines the proportion of heat energy that is transformed into power. More precisely, it is defined as power output divided by heat input usually expressed as a percentage. The second law of thermodynamics puts a fundamental limit on the thermal efficiency of heat engines, such that, surprisingly, even an ideal, frictionless engine cannot convert anywhere near 100% of its input heat into useful work. The limiting factors are the temperature at which the heat enters the engine, T_H, and the temperature of the environment into which the engine exhausts its waste heat, T_C, measured in Kelvin, the absolute temperature scale. For any engine working between these two temperatures [8] Carnot's theorem gives: thermal efficiency [8] is equal to or less than one minus the ratio of T_C/T_H. In essence what this means is that we cannot extract more heat from the steam or fuel than is permitted by the dictates of the second law and the requirements of entropy. For T_C to be lower than the ambient temperature we would be requiring a lowering of entropy. Ice, for example, has less entropy than warm water vapour at room temperature. However, it requires power input to form ice as we know from the cost of refrigeration, unless we live in Antarctica! The limiting value is called the Carnot cycle efficiency because it is the efficiency of an ideal, lossless (reversible) engine cycle, termed the Carnot cycle. The relationship essentially states that no heat engine, regardless of its construction, can exceed this efficiency.

Examples of T_H are the temperature of hot steam entering the turbine of a steam power plant, or the temperature at which the fuel burns in an internal combustion engine. T_C is usually the ambient temperature where the engine is located, or the temperature of a lake or river into which waste heat is discharged. For example, if an automobile engine burns gasoline at a temperature of $T_H = 1{,}500°\text{F} = 1{,}089$ K and the ambient temperature is $T_C = 70°\text{F} = 294$ K, then its maximum possible efficiency [8] is

$$\eta_{\text{th}} = \left[1 - \frac{294}{1{,}089}\right] \times 100 = 73\%$$

Real automobile engines are much less efficient than the maximum possible efficiency at only around 25%. Combined cycle power stations have efficiencies that are considerably higher but will still fall at least 15 percentage points short of the Carnot value. A large coal-fuelled electrical generating plant turbine peaks at about 36%, whereas in a combined cycle plant thermal efficiencies approach 60%.

In converting thermal energy into electrical power, we can therefore conclude that a conventional modern power plant in which a steam turbine drives a synchronous generator, the conversion efficiency will be at best $0.6 \times 0.9 \times 100 = 54\%$ in a power station operating in combined cycle mode where waste heat is used to warm the houses of the local community. A coal-fired power station that delivers 500 MW of electrical power to the grid dissipates almost the same amount in the form of heat. If you have given some thought to thermodynamics and the second law, you will not be surprised to learn that this just leaks inexorably into the atmosphere!

1.4 Summary

Once the era of renewables is established, it will be one in which high-speed global travel by air will be essentially impossible for the large bulk of humanity because viable fuels such as kerosene or hydrogen will be scarce and unaffordable for large-scale transport of people and goods. The widespread movement of cargo will probably be limited to solar and wind-powered container ships, or to trans-continental electrified railways. The transport of goods over long distances is likely to become prohibitively expensive in an energy-constrained world. Markets will surely become predominantly regional and local, as a result of the disappearance of the primary *modus operandi* for many businesses and companies of global reach, which have relied on cheap Third World labour, backed up by inexpensive and rapid transport, to undercut local production. In this new age, power will be supplied almost wholly through electrical sockets and it will be limited and expensive. So it seems reasonable to presume that there will be prolonged and intensive drives to improve efficiency in electrical machines – the primary source of motive power – not because of the desire for pollution reduction or prevention that the renewable power generators will satisfy, but by the need to maximise access to limited and expensive electricity by fully optimising its use.

Many of the described inefficiencies in previous sections are unavoidable in electrical equipment and consumer goods that incorporate motors and transformers – the main exception being battery-operated devices, unless the batteries are rechargeable in which case the charger will waste energy in the form of heat generated in the transformer. A particular example of ineptitude in electrical power distribution is found in many, if not most, huge office blocks housing hundreds of mains-driven computers all with their own energy-inefficient transformers and rectifiers. All such premises should have a single AC/DC conversion unit that supplies all computers with the requisite DC voltages. This kind of power profligacy is not just confined to call centres and administrative offices but is to be found in all 'walks of life' from homes to industry. The role that can be played by the application of electromagnetic levitation and suspension techniques, including super-conduction methods, to the raising of efficiency throughout the electrical industries, will be explored in succeeding chapters.

1.5 References

[1] Gregory, M.S. *History and development of engineering*. London: Longman Group Ltd.; 1971

[2] Kirkby, R.S., Withington, S., Darling, A.R., & Kilgour, F.G. *Engineering in history*. London: McGraw-Hill Book Co.; 1956

[3] *UK energy review*. Available from http://www.berr.gov.uk/files/file32003.pdf

[4] Sangster, A.J. *Energy for a warming world*. London: Springer-Verlag Ltd.; 2010

[5] World Energy Consumption. Available from www.solcomhouse.com/worldenergy.htm

[6] McKenzie Smith, I. *Hughes electrical technology*. Harlow, UK: Prentice Hall; 1995
There is a multitude of text books that provide comprehensive instruction on electrical machines. This one keeps it simple.
[7] Chapman, S.J. *Electric machinery fundamentals*. New York: McGraw-Hill International; 1985
This text provides a very detailed examination of the topic of electrical machines. It is aimed at second/third-year undergraduates at a UK university
[8] Howell, J.R., & Buckius, R.D. *Fundamental of engineering thermodynamics*. New York, NY: McGraw-Hill International; 1987

Chapter 2
Electromagnetic fundamentals

In atomic theory we have fields and we have particles. The fields and the particles are not two different things. They are two ways of describing the same thing – two different points of view.

P.A.M. Dirac

2.1 Basic electromagnetic theory

2.1.1 Maxwell's equations

In traditional electrical engineering science, at the macroscopic level, all electrical phenomena can be interpreted as being evolved from the forces acting between still, or moving, 'point' charges (electrons and protons). In fact, four concise equations, commonly referred to as the Maxwell equations, are sufficient to describe all known macroscopic field interactions in electrical science. These equations are, in a minimalist mathematical form:

$$\text{Flux } \mathbf{D} = \text{charge enclosed} \tag{2.1}$$

$$\text{Flux } \mathbf{B} = 0 \tag{2.2}$$

$$\text{Circ } \mathbf{H} = I + \text{rate of change of electric flux} \tag{2.3}$$

$$\text{Circ } \mathbf{E} = \text{rate of change of magnetic flux} \tag{2.4}$$

At first sight these equations can seem perplexing and perhaps incomprehensible, but once the symbols and the terminology are established, and the historical development is explained, the obscurity should disappear. To this end, the following symbol identifications and further definitions are appropriate.

- Electric charge (Q), which may be either positive or negative, is conserved in all electrical operations.
- The electric current through any surface is the rate at which charge passes through the surface: that is $I = \rho v A$, where ρ is the charge density in C/m^3, v is the velocity of moving charge in m/s, while A is the area in m^2 through which charge is passing. The dimensions of I are C/s, which is an Ampere in the metre.kilogram.second (m.k.s.) system.

- The electric current through any *closed* surface is minus the rate of change of the charge enclosed within the surface ($I = -\mathrm{d}Q/\mathrm{d}t$). This is a general statement of Kirchoff's law, which for circuit engineers mainly appears in the more familiar form $\Sigma I = 0$ at a network junction.

For engineering purposes the basic carrier of charge, namely the electron, is considered to be a particle, essentially because the electron wavefunction in quantum mechanics displays an extremely small wavelength, $\lambda_e = 0.165$ nm. This is too short to be detected by engineering instruments. Consequently, the electron's wave-like behaviour is rarely encountered in engineering applications. Ironically, however, a macroscopic example of the need to recognise the nature of an electron is presented by superconduction, which is very pertinent to electromagnetic levitation, as we will see. It manifests itself, in particular, in the superconducting ring, which is discussed in Chapter 6.

In (2.1)–(2.4) the vector quantities **E** and **B** (denoted by bold type) are the fundamental electric and magnetic field quantities in electromagnetism, while **D** and **H** are auxiliary fields. Materials embedded within electrical systems are defined electrically by three parameters, namely conductivity (σ, mhos/m), permittivity (ε, F/m) and permeability (μ, H/m). All of these quantities are defined and dimensioned more comprehensively in reference 1.

2.1.2 *Flux and circulation*

The flux and circulation integrals embedded in (2.1)–(2.4) can be further defined in a relatively non-mathematical form, if averaging (in essence integration) can be considered to be a process that is not unduly remote from 'common sense' (see [2] and [3] and fig. 2.1). Consequently, we have for an arbitrary vector **A**:

$$\begin{aligned}\text{Flux } \mathbf{A} &= \text{average normal component of } \mathbf{A} \text{ over a} \\ &\quad \text{surface area } \mathbf{dS} \ (A_n \text{ say}) \text{ multiplied} \\ &\quad \text{by area } \mathrm{d}S \\ &= A_n \times \mathrm{d}S\end{aligned} \tag{2.5}$$

The introduction of vector algebra in (2.5) permits a 'short hand' representation of the process. For an infinitesimally small area (dS), which can be considered to be directionally aligned with a unit vector **n** normal to its surface, then a simple dot-product gives:

$$\textbf{Flux A} = \mathbf{A} \cdot \mathbf{dS} \tag{2.6}$$

For a surface area S of finite size, we then have (see [2], [4]):

$$\textbf{Flux A} = \sum_S \mathbf{A} \cdot \mathbf{dS} = \iint_S \mathbf{A} \cdot \mathbf{dS} \tag{2.7}$$

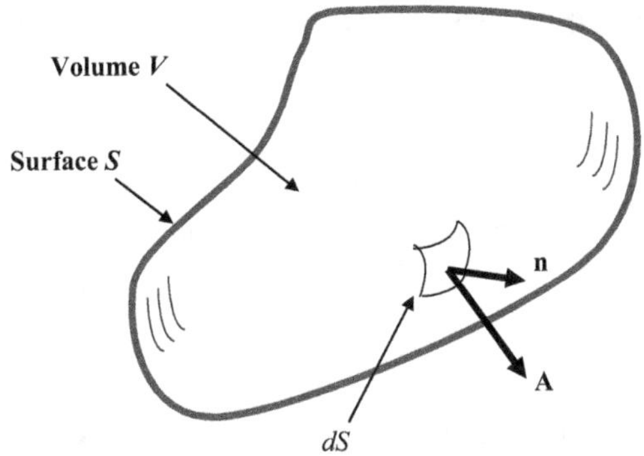

Figure 2.1 The closed surface S defines the volume V. The direction of the elemental surface dA is defined by the unit vector **n**, *and vector* **A** *represents in magnitude and direction the field passing through it*

If the surface of interest is not open, as above, but closed like the surface of a balloon, then (2.7) takes the form:

$$\textbf{Flux A} = \oiint_S \mathbf{A} \cdot \mathbf{dS} \tag{2.8}$$

The mathematical form of circulation can be constructed in a similar manner from the basic definition:

$$\begin{aligned} \text{Circ } \mathbf{A} &= \text{average tangential component of } \mathbf{A} \text{ along} \\ &\quad \text{path } \mathbf{d}\boldsymbol{\ell}\ (A_t \text{ say) times the length of path } d\ell \\ &= A_t \times d\ell \end{aligned} \tag{2.9}$$

In vector notation the 'circulation' (or perhaps it should be 'translation' for an open path) along the elemental path $\mathbf{d}\boldsymbol{\ell}$ is given by (Figure 2.2)

$$\text{Circ } \mathbf{A} = \mathbf{A} \cdot \mathbf{d}\boldsymbol{\ell} \tag{2.10}$$

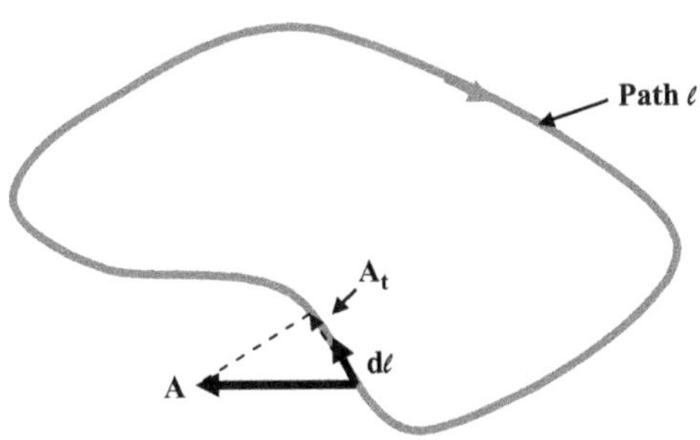

Figure 2.2 Circulation of **A** *around the path ℓ is the line integral of the tangential component A_t*

For an arbitrary path of length ℓ circulation is expressed mathematically in the form:

$$\text{Circ}\ \mathbf{A} = \int_{\ell} \mathbf{A} \cdot \mathbf{d}\ell \tag{2.11}$$

For a closed path or loop, which is much more common in electrical calculations, we get:

$$\text{Circ}\ \mathbf{A} = \oint_{\ell} \mathbf{A} \cdot \mathbf{d}\ell \tag{2.12}$$

With the above vector definitions in place, we can now express the Maxwell equations in their vector integral form as follows:

$$\oiint_S \mathbf{D} \cdot \mathbf{dS} = Q_{\text{free}} \tag{2.13}$$

$$\oiint_S \mathbf{B} \cdot \mathbf{dS} = 0 \tag{2.14}$$

$$\oint_C \mathbf{H} \cdot \mathbf{d}\ell = I_{\text{cond}} + \frac{\partial}{\partial t} \iint_A \mathbf{D} \cdot \mathbf{dA} \tag{2.15}$$

$$\oint_C \mathbf{E} \cdot \mathbf{d}\ell = -\frac{\partial}{\partial t} \iint_A \mathbf{B} \cdot \mathbf{dA} \tag{2.16}$$

In (2.13) Q_{free} denotes the free, unbounded charge with the closed surface S, while in (2.15) I_{cond} denotes the conducting current, or free charge passing through the open surface A that spans the circuital path C. The second term on the right of (2.15) is Maxwell's displacement current that also threads through the surface A.

2.1.3 Boundary conditions

At a 'smooth' interface between two different materials (1 and 2), where smooth implies surface roughness features which are very much less than the free-space wavelength at the frequency of interest, (2.13)–(2.16) reduce to the following boundary conditions:

$$\mathbf{n} \cdot \mathbf{D}_1 = \mathbf{n} \cdot \mathbf{D}_2 \tag{2.17}$$

$$\mathbf{n} \cdot \mathbf{B}_1 = \mathbf{n} \cdot \mathbf{B}_2 \tag{2.18}$$

$$\mathbf{n} \times \mathbf{E}_1 = \mathbf{n} \times \mathbf{E}_2 \tag{2.19}$$

$$\mathbf{n} \times \mathbf{B}_1 = \mathbf{n} \times \mathbf{B}_2 \tag{2.20}$$

If material 2 is a 'good conductor' the following forms apply:

$$\mathbf{n} \cdot \mathbf{D}_1 = \rho_s \tag{2.21}$$

$$\mathbf{n} \cdot \mathbf{B}_1 = 0 \tag{2.22}$$

$$\mathbf{n} \times \mathbf{E}_1 = 0 \tag{2.23}$$

$$\mathbf{n} \times \mathbf{H}_1 = \mathbf{J}_s \tag{2.24}$$

where **n** is the unit normal to the surface, ρ_s is the charge density on the surface and $\mathbf{J}_s$ is the surface current density.

In electromagnetism the fundamental force equation attributed to Lorentz [2] can be expressed in vectorial form as follows:

$$[\mathbf{F} = Q(\mathbf{E} + \mathbf{u} \times \mathbf{B})] \tag{2.25}$$

In the m.k.s. system, we already know that force is expressed in Newtons, Q is measured in Coulombs and velocity **u** in metres/second. In this system, therefore, electric field has the dimension Newton/coulomb (N/C) while magnetic flux density **B** has the dimension Ns/mC. Needless to say, we do not use these clumsy forms. In the m.k.s. system electric field has the basic dimension Volt/meter (V/m) while magnetic flux density gets the dimension Tesla (T). The relationship between a V/m and an N/C and between a Tesla and an Ns/mC can be found in reference 1.

2.1.4 Electrostatics

If (2.13) is applied to a small spherical volume of stationary and isolated free charge (q), located in a magnetic field-free region, then symmetry and the circulation law for E (2.16) dictate that the electric vector **D** will be everywhere radial and unvarying over an enveloping spherical surface (S) of radius r (Figure 2.3). Under these circumstances

$$\oiint_S \mathbf{D} \cdot \mathbf{dA} = 4\pi r^2 D_r$$

Clearly, Maxwell's first equation reduces to the inverse square law, or Gauss' law that is usually written as

$$\left[E_r = \frac{q}{4\pi r^2 \varepsilon_0}\right] \tag{2.26}$$

provided that, for air, or vacuum

$$\mathbf{D} = \varepsilon_0 \mathbf{E} \tag{2.27}$$

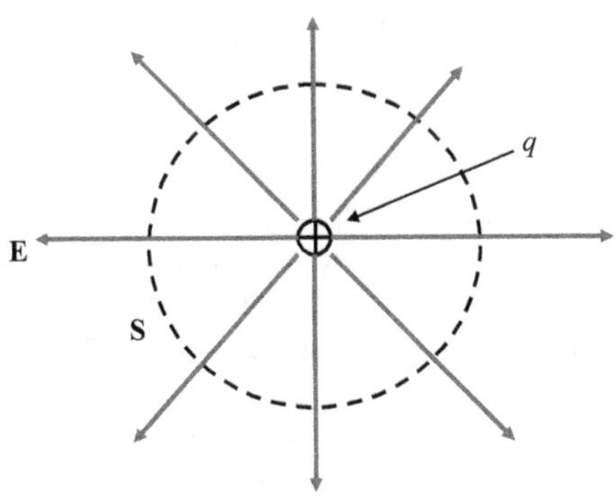

Figure 2.3 Electric field flux through a spherical surface containing a point charge q

In the m.k.s. system, **D** has the dimension C/m^2, while the constant of proportionality ε_0 is termed the absolute permittivity of free space. Measurement reveals that it has the value 8.84×10^{-12} F/m, where in the m.k.s. system a Farad is a Coulomb/volt. More generally, $\mathbf{D} = \boldsymbol{\varepsilon}_r\varepsilon_0\mathbf{E}$, where $\boldsymbol{\varepsilon}_r$ is termed the relative permittivity of the material concerned. It need not be a constant, and could be a complex tensor to represent a general non-linear, anisotropic, inhomogeneous medium.

In conjunction with (2.25), (2.26) yields the Coulomb force law that can be expressed as

$$\left[F = \frac{qQ}{4\pi r^2 \varepsilon_0}\right] \quad \text{Newtons} \tag{2.28}$$

for isolated point charges q and Q (in Coulombs) distance r metres apart in a vacuum.

2.1.5 Magnetostatics

The fundamental measurement-based laws of magnetostatics, as deduced by the early pioneers, can be constructed in similar manner from (2.14) and (2.16). Equation (2.14), usually referred to as the flux law for magnetic field, simply states that however generated – by a permanent magnet, by the Earth's core, by current on a wire, by a current carrying coil – magnet flux density forms closed loops. At the macroscopic level there are no magnetic field sources performing the same role as charge in electrostatics. Hence, if we apply (2.16) to a steady (DC) current (I) on a long straight wire, we can confidently say that the magnetic field that the law predicts must form closed circular loops centred on the wire – assuming the wire has a cylindrical cross-section. We therefore obtain:

$$\oint_C \mathbf{H} \cdot \mathrm{d}\boldsymbol{\ell} = 2\pi r H_t = I$$

$$H_t = \frac{I}{2\pi r} \tag{2.29}$$

Equation (2.29) is Ampere's law, and since I is expressed in amperes (A) in m.k.s. system, the dimension of magnetic field strength H is A/m. With H in A/m and B in Tesla, consistency of the magnetic field laws requires that for vacuum

$$\mathbf{B} = \mu_0 \mathbf{H} \tag{2.30}$$

where the constant μ_0 is termed the absolute permeability of free space and has the magnitude $4\pi \times 10^{-7}$ Henries/m (Henry = Tm^2/A). As with electrostatics we can say that more generally, $\mathbf{B} = \mu_r\mu_0\mathbf{H}$, where μ_r is termed the relative permeability of the material concerned. It need not be a constant, being a function of **H** in a ferrite material, and it could be a complex tensor to represent a general, anisotropic, inhomogeneous medium. Application of the circuital law for H to a current element $I\,\mathrm{d}\ell$ rather than to a current on a long straight wire, leads to the Biot and Savart law (see reference 1), which is usually expressed in the form:

$$\left[\mathrm{d}B = \frac{\mu_0 I\,\mathrm{d}\ell}{4\pi r^2}\right] \tag{2.31}$$

For completeness, it should be noted that in addition to permittivity and permeability materials also exhibit an ability for conduct free charge, some much better than others. This facility is expressed through the conductivity parameter s (mhos/m), which features in Ohms law. In field terms, Ohms law has the form:

$$\mathbf{J} = \sigma \mathbf{E} = \frac{\mathbf{E}}{\rho} \tag{2.32}$$

where **J** is current density in A/m^2 and ρ is resistivity in ohms.m. Equations (2.27), (2.30) and (2.32) are termed the constitutive equations and provide the means of incorporating the macroscopic characteristics of real materials into theoretical electrical science and in particular electromagnetic wave theory.

2.2 Plane wave solution

In time varying electrical devices and systems in which electric and magnetic fields exist in combination, electromagnetic waves are unavoidable. These can appear in quite complex trapped, surface, evanescent and radiant embodiments. In these circumstances the integral forms of Maxwell's equations, developed above, become inappropriate since the finite volumes, surfaces and paths over which integrations have to be performed are no longer identifiable. What is required in this case is a set of equations that represent the field behaviour at a point in space. The conversion from the integral forms to these point forms (differential forms) of Maxwell's equations is developed in most textbooks on the topic [1–8] and essentially entails the recruitment of well-known vector–differential theorems such as the divergence theorem and Stokes' law to affect the transitions.

If we limit our examination to field theory problems of the propagation type, for example electromagnetic wave propagation between parallel conducting plates and transverse electromagnetic wave (TEM) propagation in the atmosphere, as is appropriate for levitation applications, then we need to consider only source-free problems. These are so described, because in problems of the above description, the region of interest can be declared to be remote from the sources producing the fields, and the problem reduces to a boundary value problem, for which Maxwell's equations, in differential form, become

$$\nabla \cdot \mathbf{D} = 0 \tag{2.33}$$

$$\nabla \cdot \mathbf{B} = 0 \tag{2.34}$$

$$\nabla \times \mathbf{E} = -\frac{\partial \mathbf{B}}{\partial t} \tag{2.35}$$

$$\nabla \times \mathbf{H} = \frac{\partial \mathbf{D}}{\partial t} \tag{2.36}$$

where **E** and **H** represent the electric and magnetic field intensities in the region of interest. As before, $\mathbf{D} = \varepsilon\mathbf{E}$ is the electric flux density while $\mathbf{B} = \mu\mathbf{H}$ is the magnetic flux density. The 'del' operator (∇) expresses directional derivatives in the three space directions. It is a vector, which in the Cartesian system (for example) has the form:

$$\nabla = \mathbf{i}\frac{\partial}{\partial \mathbf{x}} + \mathbf{j}\frac{\partial}{\partial \mathbf{y}} + \mathbf{k}\frac{\partial}{\partial \mathbf{z}} \tag{2.37}$$

When the del operator is multiplied to a scalar ($\phi(x, y, z)$ say) the result is a vector that expresses the gradient or slope of the function ϕ in all three space directions, that is

$$\nabla\phi = \mathbf{i}\frac{\partial\phi}{\partial \mathbf{x}} + \mathbf{j}\frac{\partial\phi}{\partial \mathbf{y}} + \mathbf{k}\frac{\partial\phi}{\partial \mathbf{z}} \tag{2.38}$$

Cross-multiplication of del with a vector produces the operation of 'curl', while dot multiplication produces the operation of 'divergence' ('div'). Crudely, curl is circulation at a point, while divergence is flux at a point.

2.2.1 Second-order differential equation

To solve Maxwell's equations for the electric field components or the magnetic field components in a bounded region it is first necessary to form an equation in either E or H alone. The standard procedure for achieving this conversion is to perform a curl operation on either the curl equation for E or the corresponding equation for H. This gives, for example, using (2.35)

$$\begin{aligned} \nabla \times \nabla \times \mathbf{E} &= -\frac{\partial}{\partial t}\mu\nabla \times \mathbf{H} \\ &= -\frac{\partial}{\partial t}\left[\frac{\partial}{\partial t}\mu\varepsilon\mathbf{E}\right] \\ &= -\mu\varepsilon\frac{\partial^2 \mathbf{E}}{\partial \mathbf{t}^2} \end{aligned} \tag{2.39}$$

Hence, on using a convenient vector identity, which states that for any vector $\mathbf{A}$

$$\nabla \times \nabla \times \mathbf{A} = \nabla\nabla \cdot \mathbf{A} - \nabla^2\mathbf{A} \tag{2.40}$$

Equation (2.39) can be re-expressed as follows:

$$\nabla\nabla \cdot \mathbf{E} - \nabla^2\mathbf{E} = -\mu\varepsilon\frac{\partial^2 \mathbf{E}}{\partial \mathbf{t}^2} \tag{2.41}$$

But, from (2.33), $\nabla \cdot \mathbf{E} = 0$, for a linear, homogeneous medium for which μ and ε are constants. Therefore

$$\nabla^2\mathbf{E} = \mu\varepsilon\frac{\partial^2 \mathbf{E}}{\partial \mathbf{t}^2} \tag{2.42}$$

and by analogy

$$\nabla^2\mathbf{H} = \mu\varepsilon\frac{\partial^2 \mathbf{H}}{\partial \mathbf{t}^2} \tag{2.43}$$

Equations (2.42) and (2.43) are wave equations. Equations of this nature, with appropriate variables, appear in most branches of science and engineering and their solutions have been studied widely. Solutions depend very much on the boundary conditions – namely the conditions imposed on the variables at the periphery of the solution region. They can fix the magnitude of the variable (Dirichlet condition) or the rate of change of the variable (Newman condition) or a mixture of both. The unique solution depends on the conditions being neither underspecified nor overspecified.

For example, let us consider formulating a solution to (2.42), and inevitably (2.43) because of the Maxwell linkages, for a region of space ($\mu = \mu_0 : \varepsilon = \varepsilon_0$) that is large enough to presume that all boundaries are effectively at infinity. In this case we can choose to represent the region mathematically using Cartesian coordinates, and furthermore since we anticipate that the solution is a waveform we can arbitrarily determine that the waves travel in the z-direction. This implies that the rates of change of the E-field in x and y are zero, and using (2.33) it follows that $E_z = 0$. The equation to be solved, therefore, is

$$\frac{\partial^2 \mathbf{E}}{\partial \mathbf{z}^2} = \mu_0 \varepsilon_0 \frac{\partial^2 \mathbf{E}}{\partial \mathbf{t}^2} \tag{2.44}$$

where, in general, $\mathbf{E} = \boldsymbol{i}E_x + \boldsymbol{j}E_y$. However, if we choose to align the coordinate system so that $\mathbf{E}$ lies along the x-axis (x-polarised solution), then $E_y = 0$ and the wave equation reduces to the scalar form:

$$\frac{\partial^2 E_x}{\partial \mathbf{z}^2} = \frac{1}{c^2} \frac{\partial^2 E_x}{\partial \mathbf{t}^2} \tag{2.45}$$

if $c = \frac{1}{\sqrt{\mu_0 \varepsilon_0}}$.

2.2.2 General solution

Equation (2.17) has a wave solution of the general form:

$$E_x = Af(z - ct) + Bf(z + ct) \tag{2.46}$$

This is easily demonstrated by substitution back into the equation. The first term represents a wave travelling in the $+z$-direction while the second allows for a reflected wave, if such exists. Given that velocity is the rate of change of z with respect to time, it is evident that c represents velocity (actually phase velocity) of the electromagnetic wave in 'free space'. For vacuum it is equal to 3×10^8 m/s. The application of Maxwell's equations also gives $H_z = 0$ and

$$H_y = \frac{A}{\eta} f(z - ct) + \frac{B}{\eta} f(z + ct) \tag{2.47}$$

Also

$$\frac{E_x}{H_y} = \pm\sqrt{\frac{\mu_0}{\varepsilon_0}} = \pm\eta \tag{2.48}$$

where η is termed the free-space wave impedance, which for air or vacuum has the value $120\pi\,\Omega$. The resultant solution is a plane electromagnetic wave, also termed a transverse electromagnetic (TEM) wave, for which E and H are transverse to the direction of propagation and orthogonal to each other. E and H are also in time phase, as (2.48) attests. Electrical engineers are generally very familiar with the relationship between power (P), voltage (V) and current (I) in the form:

$$P = \frac{1}{2} VI \quad \text{Watts} \tag{2.49}$$

where V and I are defined in peak, rather than in the more common r.m.s., format. But, as we have seen from electrostatics, voltage is simply integrated electric field E (V/m), and from Ampere, current is integrated magnetic field intensity H (A/m), so by analogy we can suggest that for the plane wave

$$p = \frac{1}{2} EH \quad \text{W/m}^2 \tag{2.50}$$

This means that p is the real power flow density in the TEM wave. In general, complex power flow density in an electromagnetic wave is given by the Poynting vector **S**, where

$$\mathbf{S} = \frac{1}{2} \mathbf{E} \times \mathbf{H} \quad \text{W/m}^2 \tag{2.51}$$

In electrical engineering it is much more usual to examine wave solutions at a single frequency (ω, rad/s) – namely sinusoidal solutions. This actually incurs little loss of generality, since any arbitrary time variation carried on a radio wave can be resolved into a spectrum of single frequency components. The adoption of a single frequency, or a spectral frequency, in carrying through time varying computations has the distinct advantage that the time variable can be omitted. The calculations are then progressed in phasor notation. In trignometric form (2.47) becomes

$$E_x = A \exp j(\omega t - \beta z) + B \exp j(\omega t + \beta z) \tag{2.52}$$

where A and B are complex constants. The phasor form is

$$E_x = |A| \exp(-j\beta z + \varphi) + |B| \exp(j\beta z + \vartheta) \tag{2.53}$$

with φ and ϑ representing the phases, respectively, of A and B.

2.3 Waveguides and resonators

An electromagnetic resonator that will be familiar to virtually all readers in the industrialised world is the microwave oven. It comprises a metal box with highly conducting walls. Such an enclosure can trap an electromagnetic 'mode' that 'rings' at the frequency of the magnetron source. There are many possible modes and many different shapes of resonator, but we will focus here on rectangular cavity resonators and on mode patterns that could assist in providing levitation.

In section 2.2, we saw that electromagnetic fields and waves in real-world situations must be solutions of the Maxwell equations and hence the wave equations ((2.42) and (2.43)), together with properly posed boundary conditions.

A range of possible cavity shapes could yield field patterns suitable for levitation, but we will focus on the rectangular shape that is relatively easy to solve and is sufficiently representative of real structures to provide meaningful insights.

However, because of the multiplicity of modes that are possible in metal enclosures it is more usual to generate the field solutions for such structures, including the rectangular box cavity, by directly solving the wave equations ((2.42) and (2.43)). This usually commences with the generation of the modal solutions for the interior of an infinitely long rectangular metal tube or waveguide with perfectly conducting walls (Figure 2.4).

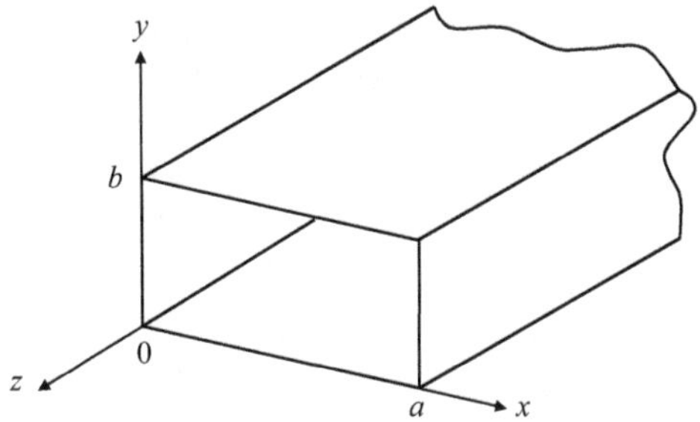

Figure 2.4 Rectangular waveguide coordinate system

It is pertinent to note, at the outset, that the solutions to Maxwell's equations can be of four distinct types. In the Cartesian frame the categories are related to the existence or otherwise of field components pointing in the propagation direction of the wave. If we take this to be the z-direction, as in Figure 2.4, then we can make the following statements:

$E_z = H_z = 0$ defines a TEM wave

$E_z \neq 0;\ H_z = 0$ transverse magnetic (TM)

$E_z = 0;\ H_z \neq 0$ transverse electric (TE)

$E_z \neq 0;\ H_z \neq 0$ hybrid mode (EH or HE)

As we have seen, the TEM solution occurs in unbounded free space. It also forms in two wire transmission systems such as coaxial line. In hollow pipes, solutions are of the transverse electric (TE) or transverse magnetic (TM) type, and we can solve for E_z or H_z in the knowledge that the transverse field components E_x, E_y, H_x and H_y can be deduced from the Maxwell relations. Consequently, we are required to solve the second-order differential equation

$$\nabla^2 H_z = \frac{1}{c^2}\frac{\partial^2 H_z}{\partial t^2} \tag{2.54}$$

for TE modes, and

$$\nabla^2 E_z = \frac{1}{c^2}\frac{\partial^2 E_z}{\partial t^2} \tag{2.55}$$

for TM modes.

2.3.1 TE solutions for rectangular waveguide

The wave equation (2.54) in a rectangular waveguide defined by means of a Cartesian coordinate frame can be expanded as follows:

$$\frac{\partial^2 H_z}{\partial x^2} + \frac{\partial^2 H_z}{\partial y^2} + \frac{\partial^2 H_z}{\partial z^2} = \frac{1}{c^2}\frac{\partial^2 H_z}{\partial t^2} \tag{2.56}$$

If we limit our horizon to seeking sinusoidal wave solutions at frequency ω (rad/s), then z-dependence of the field can be expressed as $\exp(-\bar{\gamma} z)$, where $\bar{\gamma}$ may be complex, while time dependence has the form $\exp(j\omega t)$. Hence, (2.56) becomes

$$\frac{\partial^2 H_z}{\partial x^2} + \frac{\partial^2 H_z}{\partial y^2} + k_c^2 H_z = 0 \tag{2.57}$$

where

$$k_c^2 = \bar{\gamma}^2 + \frac{\omega^2}{c^2} \tag{2.58}$$

In the Cartesian coordinate frame (Figure 2.4), a wave equation such as (2.57) can be solved by the method of separation of variables, which means that H_z can be expressed as $H_z = X(x)Y(y)$. On applying this solution scheme in conjunction with an application of appropriate boundary conditions (see section 2.2) on the waveguide walls at $x = 0$ and $x = a$, namely

$$\frac{\partial H_z}{\partial x} = 0 \tag{2.59}$$

and also at $y = 0$ and $y = b$

$$\frac{\partial H_z}{\partial y} = 0 \tag{2.60}$$

The following final solution for H_z is readily secured [5]:

$$H_{zmn} = C \cos\frac{m\pi x}{a} \cos\frac{n\pi y}{b} \exp -\bar{\gamma} z \exp j\omega t \tag{2.61}$$

In this equation, m and n are integers such that $m = 0, 1, 2, \ldots, \infty$ $(m = n \neq 0)$, and $n = 0, 1, 2, \ldots, \infty$ $(n = m \neq 0)$. This solution also requires that

$$k_c^2 = k_{cmn}^2 = \left(\frac{m\pi}{a}\right)^2 + \left(\frac{n\pi}{b}\right)^2 \tag{2.62}$$

Hence, from (2.30) we must have

$$\bar{\gamma}^2 = \left(\frac{m\pi}{a}\right)^2 + \left(\frac{n\pi}{b}\right)^2 - k^2 = k_{cmn}^2 - k^2 \tag{2.63}$$

where $k = \omega/c$. In this equation k is the free-space propagation coefficient, k_{cmn} is the cut-off coefficient, while γ is the waveguide propagation coefficient for the mnth solution or mode. The other field components of the mode are derived from H_z using the following relations derived from the Maxwell equations:

$$
\begin{aligned}
H_{xmn} &= -\frac{\bar{\gamma}_{mn}}{k_{cmn}^2}\frac{\partial H_{zmn}}{\partial x} \\
H_{ymn} &= -\frac{\bar{\gamma}_{mn}}{k_{cmn}^2}\frac{\partial H_{zmn}}{\partial y} \\
E_{xmn} &= -\frac{j\omega\mu_0}{k_{cmn}^2}\frac{\partial H_{zmn}}{\partial y} \\
E_{ymn} &= \frac{j\omega\mu_0}{k_{cmn}^2}\frac{\partial H_{zmn}}{\partial x}
\end{aligned}
\tag{2.64}
$$

It is important to observe that these modal field expressions form (in mathematical terms) a *complete* set. This means that any arbitrary field distribution in the waveguide, due, for example, to the presence of a discontinuity such as a post or an iris, can always be expressed as a combination or summation of the known modes of the empty waveguide by a process not unlike the use of Fourier series to represent non-sinusoidal waveforms.

2.3.2 Frequency characteristics of waveguide modes

For fast wave solutions in uniform metallic waveguides the propagation equation has the general form:

$$\bar{\gamma}_{mn}^2 = k_{cmn}^2 - k^2 \tag{2.65}$$

where k_{cmn} is a constant whose magnitude is governed by the size and shape of the waveguide, and k is the plane wave phase constant for the interior medium. For rectangular waveguide, we have already seen that

$$k_{cmn}^2 = \left(\frac{m\pi}{a}\right)^2 + \left(\frac{n\pi}{b}\right)^2 \tag{2.66}$$

It has other dimensionally dependant values for other waveguides, such as circular waveguide and elliptical waveguide. However, for all such waveguides the propagation equation has the same form, as expressed in (2.66).

In a lossless waveguide for which k_{cmn} is real and constant, and

$$k = \omega\sqrt{\mu_0\varepsilon_0} \tag{2.67}$$

is real and linearly related to frequency, two types of modal solution exist in waveguide. These are as follows:

1. Those for which $k > k_{cmn}$ giving

   $$\bar{\gamma} = j\bar{\beta}_{mn} \text{ where } \bar{\beta}_{mn} = (k^2 - k_{cmn}^2)^{1/2} \tag{2.68}$$

   where β_{mn} is termed the phase coefficient in rad/m for the mnth mode. The exponential multiplier in the field expression now has the form:

   $$\exp\left(-j(\bar{\beta}_{mn}z - \omega t)\right)$$

which implies that the mode in question propagates with no attenuation in a lossless waveguide.

2. Those for which $k < k_{cmn}$ giving

$$\bar{\gamma}_{mn} = \bar{\alpha}_{mn} = (k_{cmn}^2 - k^2)^{1/2} \tag{2.69}$$

where α_{mn} is termed the attenuation coefficient in nepers per metre (Np/m). The exponential multiplier in the field expressions now has the form

$$\exp(-\bar{\alpha}_{mn}z)\exp(j\omega t)$$

and this implies that the mode produces a field that attenuates in z. It does not propagate. Such a mode is said to be evanescent.

The frequency at which $k = k_{cmn}$ is termed the cut-off frequency. It is the frequency at which, in a lossless waveguide, a given mode changes abruptly from propagating to non-propagating. For any waveguide there is in general a separate and distinct cut-off frequency for every TE mode, and every TM mode, in the mode set. However, it is not uncommon for a TE mode and a TM mode to have the same cut-off frequency. Such modes are said to be degenerate. For a TE mode and a TM mode in rectangular waveguide the cut-off frequency is evaluated as follows:

Let ω_c denote the frequency at which $k = k_{cmn}$. Then

$$\omega_c\sqrt{\mu_0\varepsilon} = \sqrt{\left(\frac{m\pi}{a}\right)^2 + \left(\frac{n\pi}{b}\right)^2}$$

$$\text{i.e.} \qquad \omega_c = c\sqrt{\left(\frac{m\pi}{a}\right)^2 + \left(\frac{n\pi}{b}\right)^2}$$

Hence,

$$\begin{aligned} f_{cmn} &= \frac{c}{2}\sqrt{\left(\frac{m^2}{a^2} + \frac{n^2}{b^2}\right)} \\ &= \frac{c}{2ab}\sqrt{(m^2b^2 + n^2a^2)} \end{aligned} \tag{2.70}$$

Also, since $c = f\lambda$, then

$$\lambda_{cmn} = \frac{2ab}{\sqrt{(m^2b^2 + n^2a^2)}} \tag{2.71}$$

2.3.3 TE_{10} mode

The mode with the lowest cut-off frequency, which means that it can exist alone in a smooth empty waveguide, is the TE_{10} mode. The field expressions for this mode (in magnitude), using (2.61) and (2.64), are

$$\begin{aligned} H_z &= \cos\frac{\pi x}{a} \\ H_y &= 0 \\ H_x &= j\frac{\bar{\beta}_{10}a}{\pi}\sin\frac{\pi x}{a} \\ E_x &= 0 \\ E_y &= -j\frac{ka}{\pi}\eta_0\sin\frac{\pi x}{a} \end{aligned} \tag{2.72}$$

Each component in equation (2.72) is of course multiplied by the wave term:

$$\exp(-j(\bar{\beta}_{10}z - \omega t))$$

The functional forms in (2.44) are sufficiently simple to permit the construction of field distributions in pictorial form, as shown in Figure 2.5. This is also true of the next dozen or so TE and TM modes as many texts on electromagnetic theory demonstrate [1–8].

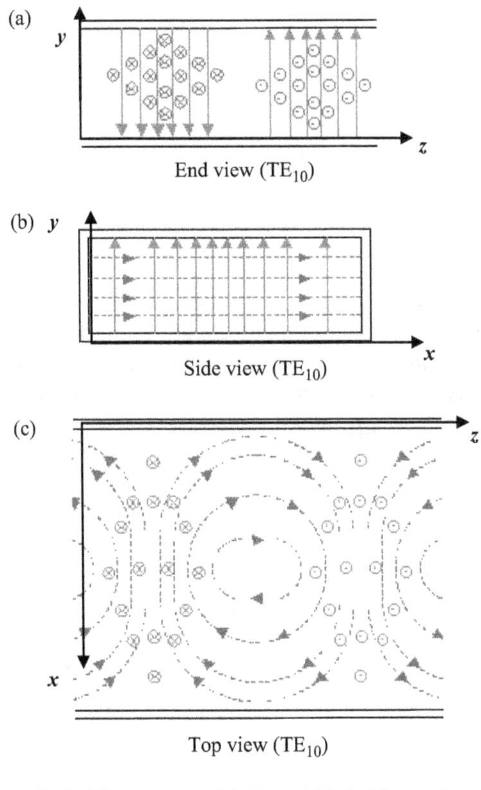

Figure 2.5 TE_{10} mode field patterns, (a) E and H-fields in the y–z plane, (b) E-field in the x–y plane, (c) E and H-fields in the x–z plane

For this TE_{10} mode the cut-off frequency is given by

$$f_{c10} = \frac{c}{2a} \quad \text{and} \quad \lambda_{c10} = 2a$$

In a typical X-band waveguide (8.0–12.4 GHz) with a broad dimension of 2 cm this simple relation sets cut-off at 7.5 GHz ($k_0a = \pi$ in Figure 2.6).

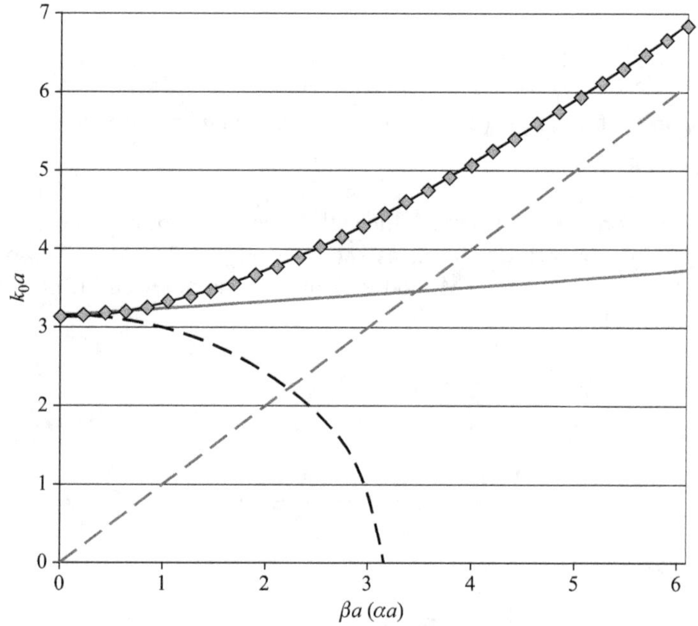

Figure 2.6 Brillouin diagram for TE_{10} mode showing frequency (k_0a) as a function of phase (βa) for rectangular waveguide (parabolic line: $a = 2$ cm, $b = 1$ cm), and as a function of αa (dashed line arc), a ring resonator (diamonds: $L = 60$ cm) and slow-wave structure with $v = 0.1c$ (inclined solid line). Grey dashed line represents the speed of light line $v = c$

A number of waveguide properties are encapsulated within the propagation equation, which are very effectively illuminated by constructing a $k_0a - \beta a$, or Brillouin diagram, for the waveguide. The beauty of this form of representation is that the slope of any line on the curve is representative of velocity, and (2.65) plots as a simple circle when $\gamma(\alpha)$ is real, and as a parabola, when $\gamma(\beta)$ is imaginary. This is shown in Figure 2.6. The figure gives a quick and comprehensive picture of the most important waveguide characteristics, such as cut-off frequencies (π for TE_{10}, 2π for TE_{20} etc.), bandwidth (distance between cut-off points on k_0a-axis), phase velocities (slope of line from origin to the frequency of interest on the TE_{10} mode propagation solution), group velocities (slope of the TE_{10} mode propagation solution at frequency of interest), higher-order mode rates of attenuation (distance from k_0a-axis to intercept on αa solution at frequency of interest), and frequency stratification of the modes. Usually only the first few TE modes are shown (TE_{10}, TE_{20}, TE_{11}, TE_{21} etc.) but in principle the complete TE_{mn} mode structure could be displayed. A similar diagram exists for TM modes. Note that the parabolic curves depicting propagating solutions lie above the v = c line (dashed-line at ~45 degrees). Such solutions for uniform waveguide are termed 'fast' waves. Below cut-off, $k_0a = \pi$ for the TE_{10} mode, the solutions are evanescent as we have already seen and these cut-off solutions are represented by the quadrant of a circle (dashed-line arc for TE_{10} mode – it is not quite circular because of the slightly different axis scales). The attenuation coefficient α in Np/m can be deduced from the horizontal distance from the frequency of interest on the k_0a-axis to the dotted curve of the mode of interest.

Consider now what happens if the waveguide is formed into a ring with the output port feeding back into the input port. The arrangement is termed a ring resonator for obvious reasons as we shall see. If the mean length of the ring is L, then in a low loss system a wave will propagate around the ring only if $p\lambda_g = L$, where p is an integer equal to or greater than unity. For the TE_{10} mode, since $\beta_{10} = \frac{2\pi}{\lambda_{g10}}$, then for resonance we must have

$$\beta_{10}a = \frac{2\pi pa}{L} \tag{2.73}$$

These solutions are depicted as blue diamonds strung along the TE_{10} mode parabola in Figure 2.6. The first diamond is the TE_{101} resonant mode, the next is the TE_{102} resonant mode and so on. The same is true for TE_{20} and higher-order waveguide modes, so that in general the ring supports a triple infinity of TE_{mnp} and TM_{mnp} resonances. Each of these resonant modes (or quanta in quantum mechanical terms) for the ring exhibits a discrete frequency, which in quantum physics would be related to an energy level. This analogy with electron wave resonances in atoms will be helpful in understanding superconductivity in Chapter 6, as we will see.

Brillouin diagrams for other types of uniform waveguide differ only in so far as the k_0a-axis intercepts are associated with the roots of functions other than trigonometric functions, for example Bessel functions of the first kind in the case of the TE and TM modes of a cylindrical waveguide. Solutions for non-smooth waveguides provide rather different Brillouin curves. For example, a waveguide periodically loaded with metal fins forms what is termed a slow-wave structure and supports hybrid HE and EH modes. The fundamental branch of such a mode is depicted in Figure 2.6 (inclined solid line). The curve lies mostly well below the v = c line indicative of a wave solution with a 'slow' phase velocity. It is also suggestive of a very narrow band structure with lower and upper cut-off frequencies. In fact periodic waveguides are pass-band filters whose nature is largely controlled by the periodicity [9]. This observation is introduced here without much in the way of elaboration in order to build an analogy between electromagnetic waveguides and electron wave behaviour in a superconductor, which, as we shall see, can be viewed as a periodic lattice of ions.

2.4 Summary

The topic of electromagnetism is extensive and deep, as the references attest. But here we have endeavoured to restrict coverage of it to a single chapter by focusing only on those aspects that are needed to illuminate later chapters.

For example, the Maxwell equations, which are presented in their classical flux and circulation formats in (2.1)–(2.4), are expanded into their integral forms in section 2.1.2 and then applied to electrostatic (section 2.1.4) and magnetostatic (section 2.1.5) cases, thus generating equations for use in Chapter 3 where elementary levitation forces are derived. The plane wave solution (section 2.2) is developed in order to furnish a framework for addressing technical issues associated with high-frequency levitation techniques in Chapters 8 and 9. Finally, in section 2.3 the nature of electromagnetic waveguides and resonators is examined in order to assist in defining superconductivity in Chapter 6 and to enable the derivation of design equations for the cavity-based levitators evaluated in Chapters 8 and 9.

2.5 References

[1] Ferrari, R. *An introduction to electromagnetic fields*. New York, NY: Van Nostrand Reinhold Co. Ltd.; 1975

[2] Johnk, C.T.A. *Engineering electromagnetic fields and waves*. New York, NY: John Wiley & Sons; 1988

[3] Feynman, R.P., Leighton, R.B., & Sands, M. *Lectures in physics, Vol. II*. London: Addison-Wesley Publishing Co.; 1972

[4] Morse, P.M., & Feshbach, H. *Methods of theoretical physics*. New York, NY: McGraw-Hill Book Co.; 1953

[5] Kraus, J.D. *Electromagnetics*. London: McGraw-Hill Book Co.; 1984

[6] Lorrain, P., & Corson, D. *Electromagnetic field and waves*. San Francisco, CA: W.H. Freeman & Co.; 1962

[7] Baden-Fuller, A.J. *Engineering electromagnetism*. New York, NY: John Wiley & Sons; 1993

[8] Hammond, P. *Applied electromagnetism*. Oxford: Pergamon Press Ltd.; 1971

[9] Bevensee, R.M. *Slow-wave structures*. New York, NY: John Wiley & Sons Inc.; 1964

Chapter 3
Invisible forces of suspension

> I maintain there is much more wonder in science than in pseudoscience. And in addition, to whatever measure this term has any meaning, science has the additional virtue, and it is not an inconsiderable one, of being true.
>
> *Carl Sagan*

3.1 A brief history of electrical science

To levitate in a vacuum an object, which is heavier than air in defiance of the force of gravity here on the surface of Earth, scientists and engineers mainly enlist forces familiar to practitioners of electrical science. These forces are represented by the electrostatic field, the magnetostatic field and electromagnetic fields (including optical manifestations). Examples of 'levitation', using acoustic pressure, have also been reported, but this is pseudo-levitation, like aerodynamic lift, since it cannot be procured without the presence of a gas. Any other forms of 'levitation' are purely in the deceived or deluded minds of the beholders. In this chapter, we will engage only with the electrical forces, whose provenance and fundamental nature will be elucidated below.

In much the same way as Newtonian dynamics can be reduced to a few basic equations based on the notions of length, mass and time, to provide a complete and consistent set of rules for motion in the solar system and beyond, so do the equations of electromagnetism [1–8] for electricity by simply adding the concept of charge to the Newtonian parameters. This choice is actually quite sensible and natural, because the scientific evidence is that charge is 'conserved', and in so being, it is very much in sympathy with energy and momentum, which feature strongly in dynamics. Energy and momentum are also conserved.

Charge, which can be either positive or negative, can also have mass. Generally however, this property is ignored since electrical forces, whether attractive or repulsive, between charges are immensely larger (10^{42} times) than gravitational attraction. The forces associated with charge are, for representational convenience, assumed to emanate from a point. Unfortunately, this leads to mathematical absurdities if the representation is carried too far. To avoid this, our 'point' is considered to be an extremely small, but finite, volume with dimensions that are negligible in comparison with measurable physical entities and also with significant electrical lengths such as wavelength or skin depth. Nuclear forces are avoided if our 'point' volume is larger than 10^{-10} m – this being of the order of the (microscopic) diameter of atoms in typical electrical materials. By restricting dimensions in classical electromagnetic theory to no less than a thousand times 10^{-10} m (that is 10^{-7} m), what we term the macroscopic range, the mathematical problem for electrical science of regarding charge as a point source is circumvented.

3.1.1 Establishing the laws

The provenance of the Maxwell equations is well established so it is merely necessary to concisely rehearse the history in this section. While electrostatic attraction and magnetic effects were known to ancient civilisations – the ancient Greeks coined the word that gave us 'electron' – it was really not until the mid-eighteenth century in Europe that the nature of electricity really began to be unravelled. With Benjamin Franklin's [1706–1790] enunciation of the law of conservation of charge in 1747 the process of discovery was initiated. Franklin's experiments on charged metallic cups containing a cork ball led Joseph Priestly [1733–1804] to suggest that, like gravitational force, electrostatic field must behave in accordance with the inverse square law. This was in ca. 1785. This law, as is indicated in Chapter 2, is subsumed into the first Maxwell equation, which is generally termed the Gauss law, after Carl Friedrich Gauss [1777–1855] who is its source. The inverse square law was essentially confirmed by Charles Augustin de Coulomb [1736–1806] by employing a torsion balance in precise (for the time) experiments designed to measure the forces between isolated, like and unlike, charges. He showed that in magnitude, this force is proportional to the product of the charge magnitudes and inversely proportional to the square of the distance between them (see Chapter 2).

Scientists in the late eighteenth century were equally intrigued by magnetism and the mysterious forces induced by currents on wires. It was realised that this effect could not be attributed to electrostatic force but was an additional force associated with charge in motion. The relationships are built into Maxwell's third law (see Chapter 2), which in turn has been evolved from the experimentally deduced magnetostatic law, relating magnetic flux density to current, due to Jean-Baptiste Biot [1774–1862] and Felix Savart [1791–1841], building on Andre-Marie Ampere's [1775–1836] law, which was first published in 1825 (see Chapter 2). At about this time Georg S. Ohm [1789–1854] discovered that a steady current flowing in a wire resulted in a voltage between the ends of the wire, and that this voltage drop was equal to a constant times the current (Ohms law: $V = IR$). The constant (R), which is material dependent, was termed resistance.

Until the early nineteenth century scientists largely believed that they were dealing with two independent disciplines – namely electrostatics and magnetostatics. However, all this changed with Hans Christian Oersted [1777–1851] who was the first scientist to observe, in 1820, the connection between electricity and magnetism (he is attributed with coining the term 'electromagnetism') and with the publication in 1832 of Michael Faraday's [1791–1867] induction experiments, and the essentially parallel developments of Joseph Henry [1797–1878] in the United states of America.

These showed that a voltage (integrated electric field) could be produced by a changing magnetic field. The law of induction (essentially Maxwell's fourth equation (2.4)) states that the voltage induced at the terminals of a coil is equal to the product of the number of turns with the rate of change of magnetic flux through the coil. The law revealed that electric and magnetic fields are inextricably linked, unless changes with time are absent (statics) – an almost impossible scenario since all matter is made up of vibrating atomic nuclei and whirling electrons. The linkage is hinted at in the electromagnetic force law attributed to Hendrik A. Lorentz [1853–1928]. It states that an isolated charge moving in a vacuum containing an electric and a magnetic field will experience a force due to the electric field, which is proportional to the charge times the electric field strength, and a force due to the magnetic field that is proportional to the product of the charge, the velocity of the charge and the magnetic flux density (see Chapter 2). However, the discipline of electromagnetism only became properly established when James Clerk Maxwell [1831–1879] published his famous treatise *Electricity and Magnetism* in 1873. In cogitating on the unification of electric and magnetic effects Maxwell must have contemplated the relationships expressed by Ampere's law, and Faraday's law of induction, and observed that they seemed to demonstrate a peculiar lack of symmetry. By adding an electric flux term to Ampere's law, Maxwell's equation (2.15) results, which states that in addition to conduction current a magnetic field can be induced by a changing electric flux, sometimes termed 'displacement current'. Full symmetry can be achieved by incorporating a 'magnetic current' term into (2.16) [8]. In adding these terms, Maxwell was able to predict the possibility of devices being devised to generate radio waves. He did this some 21 years before Oliver Lodge [1851–1940], marginally ahead of Heinrich Hertz [1857–1894], demonstrating their existence by means of an elementary spark gap transmitter.

3.1.2 Evolution of electromagnetic levitation

The historical record places the appearance of recognisable magnetic levitation to the early nineteenth century, when J. Henry noted that a current carrying wire wound on a ferrite magnet core could generate lift against the force of gravity. Curiously though, Benjamin Franklin of lightning conductor fame, apparently predicted, much earlier, that 'We may perhaps learn to deprive large masses of their gravity and give them absolute levity for the sake of ease of transport'. This often quoted remark was contained in a letter written in 1790, well before Henry's experiments. The word levity is obviously used in the sense of 'lift' but given the era when it was said, it is more likely that he was thinking of balloons or kites rather than electromagnetism.

Samuel Earnshaw [1805–1888] features strongly in the story of levitation arising from a paper he wrote in 1842 entitled 'On the nature of molecular forces which regulate the constitution of the luminiferous ether'. With such a title one could be forgiven for presuming that the contents of the paper would hardly be likely to continue to resonate today, but in fact it does. Earnshaw is credited with the uncovering of a quite profound, yet simple stability criterion based on an energy principle, first enunciated in 1744 by Pierre Louis Maupertuis (1698–1759). It is usually termed the 'Principle of

Least Action' and has been given much substance by the efforts of Joseph L. Lagrange (1736–1813) and Leonhard Euler (1707–1783). Maupertuis is quoted as observing that 'Nature is thrifty in all its actions'. By applying this principle and some rather basic physics to fields presumed to obey the inverse square law, Earnshaw deduced his stability rule. Gravitation was known to accord with the inverse square law, and Franklin had shown that electrostatic fields behave similarly. The magnetic field of an isolated pole also behaves in accordance with the inverse square law, although in practice, since isolation of a north from a south pole is not physically possible, the decay rate is somewhat faster. However, this does not invalidate the Earnshaw discovery as subsequent analysis by Werner Braunbeck, published in 1939 [9], has established. Essentially, by calculating energy potential for such fields he determined that there can be no local energy minimum, or potential well, in a region immersed in these fields, to which the corollary is that an object in suspension in inverse square law fields will be in unstable equilibrium for at least one of its six possible directions of motion (three translational directions and three rotational directions). This, of course has considerable bearing on attempts to engineer stable levitation using electric or magnetic fields. Fortunately, Earnshaw's rule applies strictly to static situations and fails for objects in motion.

The evolution of active magnetic bearings, which overcome the Earnshaw limitation, can be traced fairly easily through the patent history related to magnetic suspension. The earliest relevant patent [10] is attributed to Jesse Beams [1898–1977] of the University of Virginia in 1941. It is entitled 'Suspension of rotatable bodies' and is concerned with high-speed centrifuges with the capability of enabling the purification of the isotopes of various elements particularly in relation to nuclear technology [11]. However, the process did not really mature until advances in solid-state electronics and modern computers provided the necessary automatic control. Active magnetic bearing have now permeated into a wide range of technologies from flywheels to blood pumps, from photolithography to medicine and haptic systems.

The same active technology is also used in systems that employ linear electric drives. These are now most commonly and popularly recognised in the form of 'maglev' trains. Pioneering developments and original patents [12] in this area are attributed to Emile Bachelet [1863–1946] but the ideas contained therein were well in advance of the engineering capabilities of the period. Real progress did not materialise until Hermann Kemper [1892–1977] brought a more practical approach to the development of maglev technology [13]. It took 30 years before a full scale operational wheel-less train was properly achieved.

3.2 Magnetic levitation and suspension forces

Most people who have been fortunate enough to pass through an education system claiming to teach evidence-based science, even at a rudimentary level, will be familiar with the magnetic forces developed by a permanent magnet. Many will have been exposed to gadgets and toys that rely on magnetic repulsion or attraction

Figure 3.1 High-speed (~300 mph) maglev train (Germany)

(see Livingston [14]). However, widespread public interest in magnetic levitation was arguably seriously awakened only by the advent of magnetically levitated trains, such as the example illustrated in Figure 3.1.

The modern development of trains without wheels, lifted and guided along magnetic tracks (maglev trains), commenced in about 1965, and by the 1980s the technology had matured sufficiently to enable Japanese and German technologist to offer these high-speed systems to the market. The success of this technology is predicated on the timely development of 'high' temperature superconducting materials (see Chapter 6) together with the growing availability of an evolving integrated circuit-based computing technology, which is necessary to provide rapid sensing and effective control of the levitated vehicle. However, the design and implementation of practical systems requires a thorough appreciation of electromagnetic theory and Maxwell's equations, which facilitate the accurate modelling of magnetic circuits and magnetic fluxes and hence the levitation capabilities of any particular magnetic arrangement.

3.2.1 Lorentz force

Magnetic suspension, whether it is applied to maglev trains or to levitated bearings in motors or generators, depends primarily on utilising the Lorentz force (see Chapter 2). When applied to a long, straight conductor of length l, carrying a current I and immersed in a magnetic field B, as shown in Figure 3.2, so that l and B are orthogonally disposed, Lorentz's field law yields the following integrated form:

$$F = BIl \text{ (Newtons)} \tag{3.1}$$

In a primitive suspension set-up, we can readily compute the current I to procure levitation if the mass of the wire is known. For example, a wire of density d kg/m^3 and radius r metres will have a mass of $\pi r^2 ld$ in kilograms. Consequently, for levitation we must have

$$BIl = \pi r^2 l d g \tag{3.2}$$

where g is the gravitational acceleration (9.807 m/s^2). When applied to a 1-cm radius copper wire, embedded in a strong 1 T magnetic field, (3.2) predicts that a tiny current of slightly less than 0.03 nA is all that is needed to achieve suspension of the wire filament. However, the arrangement is completely impractical since there is no way to generate a steady current flow, no matter how small, through a 'floating', and hence, electrically isolated, wire.

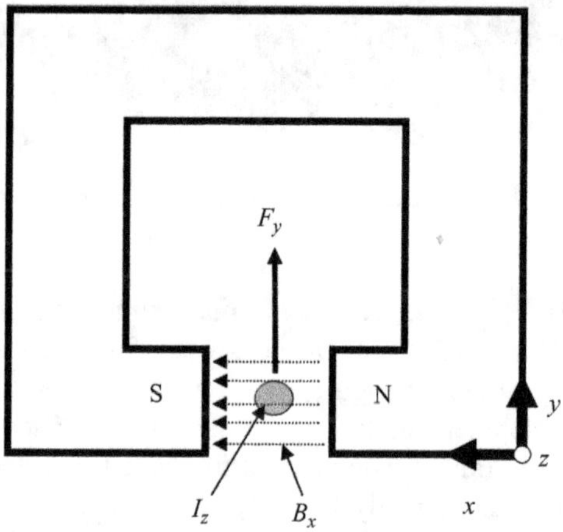

Figure 3.2 Lorentz force acting on wire of length ℓ carrying a current I_z immersed in magnetic flux of density B_x

3.2.2 Practical examples

Needless to say, practical systems for levitation are much too convoluted for a Lorentz treatment, since the active current flows are often associated with the atomic currents of orbiting and spinning electrons. Some examples are illustrated in Figure 3.3.

The permanent magnet bearing illustrated in Figure 3.3a encompasses many of the primary elements of a magnetic suspension system – a field generating fixed magnet (or coil) in close proximity to a permanent magnet 'float'. In this case, the float is attached to a guide rod to provide lateral stability. The problem with this arrangement is that if the static magnet is also a permanent magnet the separation between the float and the source magnet is uncontrolled and highly erratic, if the loading on the float should vary, as it would in a bearing application. This can be resolved by sensing the gap width, and providing a coil to enhance or decrease the source magnetic field. Such a scheme is depicted schematically in Figure 3.3b, where the static element comprises a U-shaped soft iron magnetic core with an embedded permanent magnet

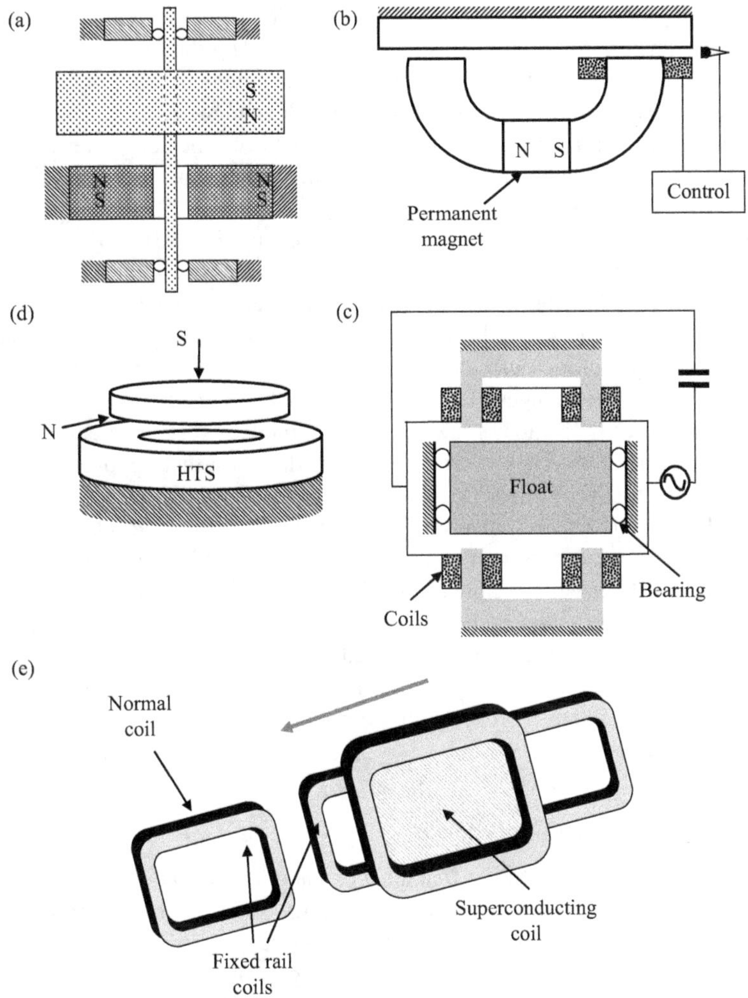

Figure 3.3 *(a–e) Some practical magnetic levitation arrangements.*
(a) Uncontrolled magnetic levitation in one dimension; (b) Magnetic levitation with simple feedback; (c) Disc magnet suspended above a superconductor; (d) Magnetic suspension system with resonant circuit control; (e) Superconducting coil levitated by movement over a fixed coil rail

field source. One leg of the U-magnet is supplied with a field enhancing coil, the current in which is adjusted in response to signals from a gap sensor. The nearer the float approaches the stator, the greater will be the coil current in a direction that enhances the repulsive force – and vice versa if the gap enlarges. The system is conditionally stable (stability is treated in more detail in Chapter 4). A physically similar scheme, using soft iron cores, permanent magnets and coils, is depicted in Figure 3.3c. It differs in that an alternating current source is used and by driving the coils through tuning capacitors, the system can be made unconditionally stable, requiring no sensors in the gaps. This is done by adjusting the capacitors so that the coil/capacitor combination is resonant at a frequency close to, but not equal to, the source frequency. So movement of the float, which changes coil inductance, drives the LC circuits closer to or further away from resonance. The coil current thus rises or falls producing suitably directed restoring forces in the gaps, thereby maintaining the float's equilibrium position.

Figure 3.3d and e depicts eddy current levitators. In Figure 3.3d, a circular annulus comprising a high temperature superconductor provides the stator for a permanent magnet disc-shaped bearing that floats above the superconductor. The vertical float position depends on the strength of the magnet. When the magnet is stationary in its equilibrium position no currents are generated in the superconducting ring. But as soon as the magnet changes position vertically, the magnetic field threading the ring changes, which by Faraday's law (Chapter 2) induces very large currents (eddy currents) in the annulus. These currents are in a direction that opposes the change (Lenz law) and produces Lorentz forces that push (or pull) the magnet back to its original position. The set-up is unconditionally stable for vertical movements. In Figure 3.3e, the above system is essentially reversed with the superconducting ring forming the float while magnet coils form the static system. Vertical stability is procured by an automatic eddy current feedback process similar to that which provides levitation in Figure 3.3d, but in addition by switching the magnetic field progressively from coil to coil along a chain of coils the float can be made to move horizontally as it maintains contact with the field. This, of course, is the essence of maglev transport systems.

3.2.3 Magnetic stored energy

By now it will be evident to readers that the direct application of the Lorentz law (3.1) to any of the above magnetic structures is verging on impossible since the fields, currents and current paths, in the magnetic cores and superconductors, are largely undefined and indeterminate, except perhaps in the case of Figure 3.3e. Consequently, if prior knowledge of levitation forces is required, without resorting to sophisticated electromagnetic simulation techniques, other less intense computational procedures have to be commandeered in order to ascertain the flotation behaviour of the 'free' element in these complex magnetic circuits.

In the examples of magnetic structures depicted in Figure 3.3, it is clear that the magnetic field, which is instrumental in producing levitation, is largely a combination of the fields generated by both the stator and the float, so a direct application of Lorentz force in order to estimate levitation strength is not possible. The solution lies in the examination of stored energy and the concept of magnetic stress. The calculation of stored energy in a magnetic field can be estimated in many ways, one of which is through the application of Ampere's law and Faraday's law (Chapter 2) to a long uniform solenoid (Figure 3.4).

If the solenoid in Figure 3.4 is long enough, it can be assumed with little error that the magnetic field generated by a current in the coil will pass predominantly

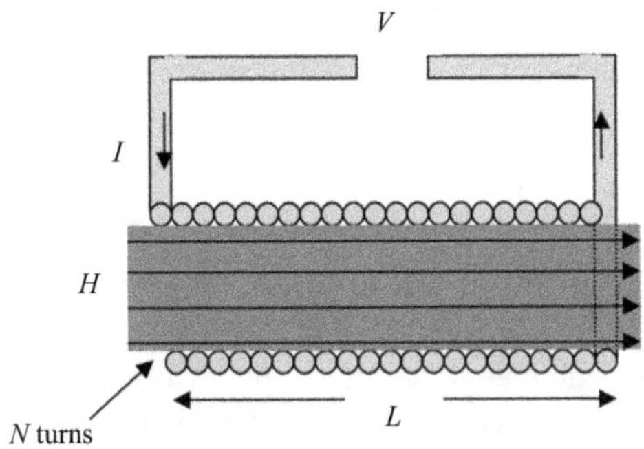

Figure 3.4 Magnetic field H in a long solenoid

through the interior of the coil with a minimal level of leakage to the exterior volume. Also the field can be presumed to be uniformly distributed over the coil cross-section. In this case, the application of Ampere's law (2.29) yields

$$H = \frac{NI}{L} \tag{3.3}$$

Hence, for an air-filled coil

$$I = \frac{BL}{\mu_0 N} \tag{3.4}$$

Now, if B is time varying, Faraday's law (2.16) leads to the result that

$$V = N\frac{\mathrm{d}\varphi}{\mathrm{d}t} \quad \text{(Volt)} \tag{3.5}$$

in which case the instantaneous power entering the terminals in the coil is given by

$$P = VI \quad \text{(Watt)} \tag{3.6}$$

Thus, the energy stored in the coil can be expressed as

$$\begin{aligned} W_\mathrm{m} &= \frac{1}{2}\int_0^T VI\,\mathrm{d}t \\ &= \frac{1}{2}\int_0^T N\frac{\mathrm{d}\varphi}{\mathrm{d}t} \times \frac{BL}{\mu_0 N}\mathrm{d}t \\ &= \frac{1}{2}BH \times \text{volume of core} \end{aligned}$$

This means that in per unit volume of the core the energy is stored to the density of

$$w_\mathrm{m} = \frac{1}{2}BH = \frac{1}{2}\frac{B^2}{\mu_0} \quad (\mathrm{J/m^3}) \tag{3.7}$$

This is actually a general expression for energy density, with μ_0 replaced by μ for materials other than air or vacuum. The most general form that accommodates inhomogeneous and anisotropic materials is

$$w_m = \frac{1}{2}\mathbf{B} \cdot \mathbf{H} \quad (\mathrm{J/m^3}) \tag{3.8}$$

3.2.4 Ferrites

While the stored magnetic energy developed by a current carrying wire or current carrying solenoid (inductor) can be readily appreciated by resorting to Ampere's law, the magnetic energy contained in a magnetic material such as iron is much more difficult to comprehend. This is because the magnetism in this case derives from quantum mechanical phenomena. It emanates from a property of electrons described as 'spin', which is rather different from spin as it is understood in classical dynamics, but still implies charge movement and a residual magnetic field in accordance with the Ampere law. The explanation of electron spin is beyond the scope of this book, but a detailed treatment of the topic can be found in references 15 and 16. From an engineering perspective, the interesting consequence of spin-generated magnetic fields in atoms is that in some metallic elements, spins and hence the associated fields can tend towards local alignment over many atoms in tiny randomly sized and shaped domains within a magnetic material (Figure 3.5a), the best known example being iron. However, when a magnetic field is applied to such a sample, domain alignment occurs (Figure 3.5b), the randomness gradually disappears and the material is said to be magnetised.

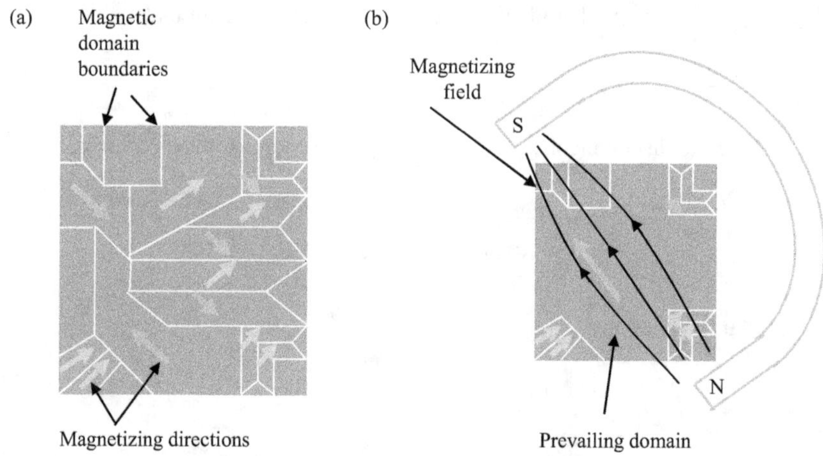

Figure 3.5 Domain representations of a magnetic material: (a) unmagnetised sample; (b) magnetised sample

In classical magnetism these domains can be viewed as tiny solenoids or tiny bar magnets of dipole moment m within the magnet as a whole (Figure 3.6). The domain 'magnets' will generally be randomly orientated in all three space directions

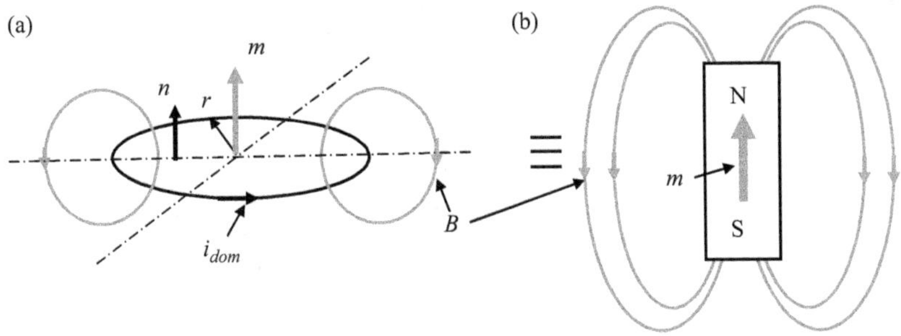

Figure 3.6 *(a) and (b) Classical bar magnet domain models. (a) Atomic current model; (b) Microscopic bar magnet model*

within an unmagnetised volume of material, as suggested, albeit in a limited 2-D form, in Figure 3.7a. When the magnetic bar is submersed in an axial magnetic field as shown, a non-linear alignment process occurs (Figure 3.7b), and the bar becomes 'permanently' magnetised (magnetisation density M).

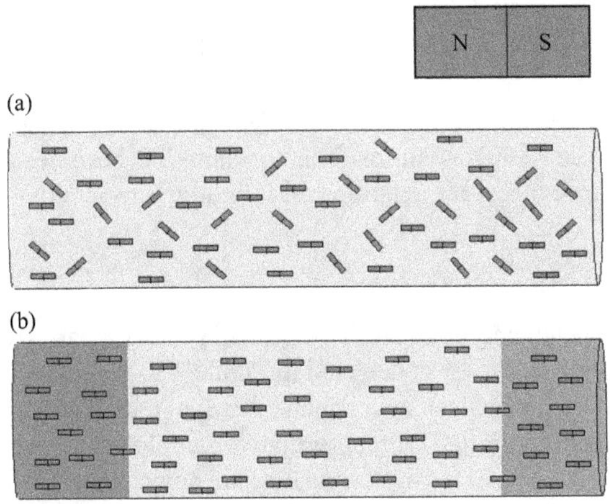

Figure 3.7 *Classical representation of magnetisation using 'domain' magnets. (a) Unmagnetised; (b) Magnetised*

An alternative classical view represents the domains by means of tiny current loops or cylinders carrying 'atomic currents' i_{dom} (Figure 3.6a) again randomly orientated in an unmagnetised ferrite bar (Figure 3.8a). The relationship between m and i_{dom} is usually expressed as

$$m = \pi r^2 i_{\mathrm{dom}} \tag{3.9}$$

where r is the radius of the cylindrical domain.

The application of a magnetising field (Figure 3.8b) leads to domain alignment, as before, with loop current axes axially directed and with sympathetic spins. This results, in effect, in a macroscopic solenoidal current of magnitude I_m. It can be obtained by integrating the aligned i_{dom} currents over the volume of the magnet.

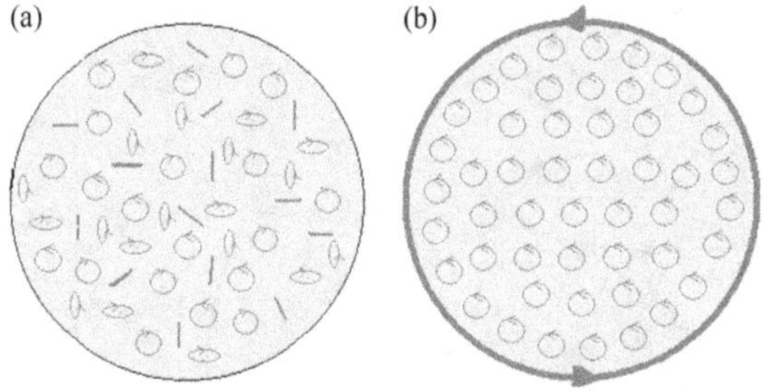

Figure 3.8 Solenoidal current representation of domains, in end view: (a) unmagnetised sample and (b) magnetisation

The result is that classically, the magnetic dipole moment per unit volume M can be equated to the macroscopic atomic current as follows:

$$M = \frac{I_{\mathrm{m}}}{L} \quad (\mathrm{A/m}) \tag{3.10}$$

where L is the axial length of the magnet in metres.

3.2.5 Hysteresis

In a ferrite material exhibiting domain alignment as it is magnetised, μ_r in the constitutive relation (see (2.30)) is no longer a constant, but becomes a function of H. Under these circumstances, the relationship is more conveniently expressed in the following form:

$$\mathbf{B} = \mu_0[\mathbf{H} + \mathbf{M}] \quad (\mathrm{Tesla}) \tag{3.11}$$

The non-linear dependence of B on H is probably most clearly represented by a hysteresis loop (or B–H curve) as depicted in Figure 3.9. These are provided as data sheets, along with their products, by magnetic material manufacturers. These are the outcome of rigorous flux testing on representative samples.

The essential elements of the hysteresis loop are depicted in the illustrative sketch provided in Figure 3.9a. It can probably be best explained by remembering that the loop is the distillation of measured results obtained by sensing the magnetic flux in a ferrite core – usually a toroidal core. If the core is initially annealed – heated to high temperature to remove any magnetism – then cooled, B and H will be zero, and the measured magnetic conditions are represented by the origin of the diagram. The introduction of an N-turn winding (Figure 3.9b) onto the core permits a magnetising force H, which is proportional to NI, to be applied to the core. This magnetising force is represented by the x-axis in the diagram (Figure 9.3a), which is usually labelled as H, but can be expressed as I or NI. Note that H is independent of the core material. As H is increased from zero to an above saturation value, the domains within the ferrite core begin to align, but slowly. The applied H field magnitude must become large enough to overcome the reluctance to align, partly as a result of friction. B grows slowly at first (dotted curve) then more quickly as an avalanching process is set up. However, once the majority of domains have been captured the

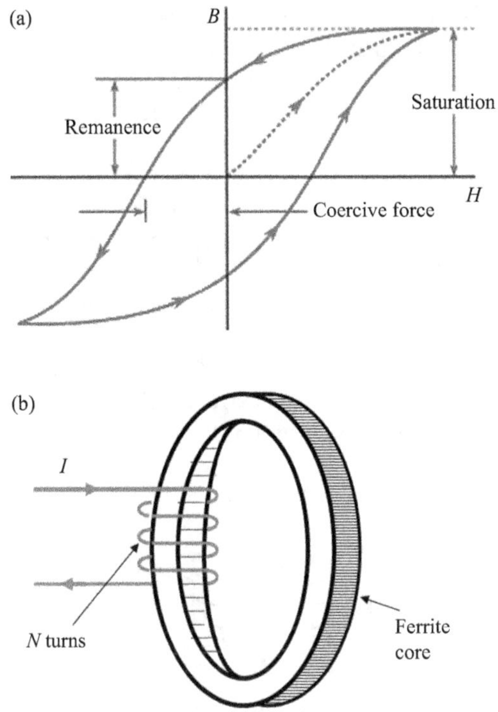

Figure 3.9 (a) Typical hysteresis loop for a ferrite material [source: © 1994 Encyclopaedia Britannica, Inc.]. (b) Ferrite toroidal core with magnetising winding

alignment process slows dramatically. The magnet is said to have reached saturation. If the current in the winding is now reduced towards zero, and subsequently increased in the opposite direction, the domains in the saturated core are initially reluctant to disalign (solid curve with left-directed arrows) until H approaches zero, whereupon rapid demagnetising occurs. At $H = 0$, the magnet is said to be in its remnant state. As H becomes increasingly negative the core loses its magnetisation totally at $B = 0$, and eventually becomes saturated in the opposite direction. The magnitude of H required to demagnetise the core is termed the coercive force. Further current reduction and reversal (solid curve with right-pointing arrows) completes a loop, which is termed the hysteresis loop. It takes real power to drive the time-varying current through the winding, and the area enclosed by the hysteresis loop is a measure of the energy lost per unit time per unit volume of the magnet, which manifests itself as heat in the core.

3.2.6 Magnetic stress

For a magnet in close proximity to another magnet, or to a current carrying superconductor, if the energy storage in the intervening space is known, then attractive or repulsive forces for levitation can be determined from differential energy calculations, associated with small displacements within the magnetic system. In magneto-mechanics, such force calculations are facilitated by the concept of magnetic stress [17,18], which is given the name Maxwell stress, although the mathematical relationships were derived by Hermann von Helmholtz (1829–1894). The attribution to Maxwell arises from his being the first to give mathematical expression to Michael Faraday's concept of the transmission of forces by means of a field. He imagined that field lines as 'lines of tension'. That stress occurs in magnets is clearly illustrated by Figure 3.7b, where the aligned domain magnets would fly apart radially, if the ferrite material were to exhibit an insufficient binding strength. The magnitude of this radial pressure can be determined by considering the solenoidal representation of a permanent magnet (Figure 3.10) depicted as a cylinder for convenience. The magnetised core is replaced by a thin cylindrical current shell $I_m = J_m \times L$, where J_m is the current density (A/m), and L is the length of the rod. This shell current produces a

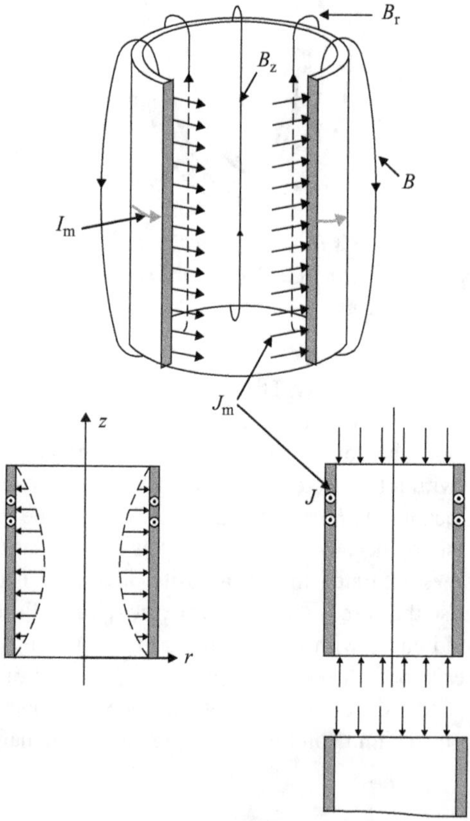

Figure 3.10 Magnetic field lines and magnetic stresses (P_r and P_z) associated with a cylindrical bar magnet (length L, radius r) represented by a current cylinder (I_m) that produces an internal axial magnetic field strength B_z

uniform internal magnetic flux density B_z, such that at the surface of the cylinder B_z and I_m are mutually orthogonal, and by Lorentz (for r.m.s. values) the combination must produce a radially directed outward force:

$$\begin{aligned} F_r &= \frac{1}{2} B_z I_m 2\pi r L \\ &= B_z H_z \pi r L \\ &= \frac{B_z^2}{\mu_0} \pi r L \quad \text{(Newtons)} \end{aligned} \tag{3.12}$$

Hence, per unit area of the cylindrical surface of the magnet, the outward pressure P_r is given by

$$P_r = \frac{B_z^2}{2\mu_0} \quad (\text{N/m}^2) \tag{3.13}$$

The repulsive end pressure on two bar magnets, on the other hand, is much more difficult to estimate using the Lorentz equation. While the shell current is known as above, the orthogonal magnetic field, which is required to produce a z-directed force, can only come from the leakage flux B_r. This field is difficult to determine accurately without full 3-D electromagnetic simulation. The solution to estimating the magnetic force in computations of this type is to examine stored energy. In the usually small gap, at least in levitation applications, between a pair of magnets, the magnetic flux density in the gap can, with little error, be assumed to be uniformly distributed in the gap (B_z). Consequently, the energy stored in the gap is

$$W_m = \pi r^2 g \frac{B_z^2}{\mu_0} \quad \text{(Joules)} \tag{3.14}$$

where g is the gap length. But this energy must be equal to the work that would have to be performed to force the poles together against the repulsive force F_z. That is

$$\begin{aligned} F_z g &= \pi r^2 g \frac{B_z^2}{2\mu_0} \\ \therefore\ F_z &= \pi r^2 \frac{B_z^2}{2\mu_0} \end{aligned}$$

and

$$P_z = \frac{B_z^2}{2\mu_0} \quad (\text{N/m}^2) \tag{3.15}$$

Clearly, pressure, in a magnetic structure relevant to levitation systems, is largely dependent on the strength of the primary magnetic flux penetrating the structure.

Equations (3.13) and (3.15) tell us that if the primary magnetic flux density within a magnetic circuit is calculable, which it usually is, then the magnetic pressure throughout the system can be estimated without difficulty. Magnetic circuit calculations are a well-established branch of magnetostatics and electromagnetic theory, and the reader is referred to reference 17 for further elaboration of the topic.

It is actually rather important that design engineers are able to perform 'ball park' calculations of this type, even in these days of highly sophisticated modelling tools. With such tools, the well-rehearsed axiom of 'rubbish in rubbish out' is by no means inappropriate, and the ability to generate prior estimates of key output variables represents a very valuable check on the performance of any simulation and on the integrity of the inputted information on which it is based.

3.3 Electric suspension force

Electrostatic suspension systems are founded on the employment of the forces that exist between electrical charges; the attractive forces between unlike charges and the repulsive forces between like charges. While the forces between point charges are calculable through the Gauss and Coulomb laws, as we have seen, finite systems formed from metallic electrodes and insulators are not susceptible to these point form laws and require a shift to Maxwell's equations (Chapter 2) to evaluate fields and forces.

3.3.1 *Energy stored*

Let us first establish the work required to produce, in the capacitor, a total charge Q_+ in the top plate (and therefore Q_- in the lower plate). q_+, q_- are intermediate values of the charge at voltage V. For a small increase in charge $\mathrm{d}q$ the work done will be $V\,\mathrm{d}q$ in, for example moving the charge to the top plate against the field; since V would be the work done if unit positive charge were to be moved to the top plate.

Thus, the work necessary to establish total charge Q_+ on the top plate is

$$W_k = \int_{q=O}^{q=Q} V\,\mathrm{d}q \quad \text{(Joules)} \tag{3.16}$$

But we have seen (Chapter 2) that by definition

$$C = \frac{\text{Charge}}{\text{Voltage}} = \frac{q}{V}$$

Therefore

$$V = \frac{q}{C}$$

Hence, we can write

$$W_k = \int_{q=Q}^{q=Q} \frac{q}{C}\mathrm{d}q = \frac{1}{2}\frac{Q^2}{C} = \frac{1}{2}VQ = \frac{1}{2}CV^2 = W_\mathrm{e} \tag{3.17}$$

where W_e is the electrostatic stored energy in Joules.

This well-known result can be expressed in field terms as follows. Consider the capacitor arrangement in Figure 3.11. As we have demonstrated in Chapter 2, for a parallel plate capacitor with a linear, homogenous, isotropic dielectric, **E** will

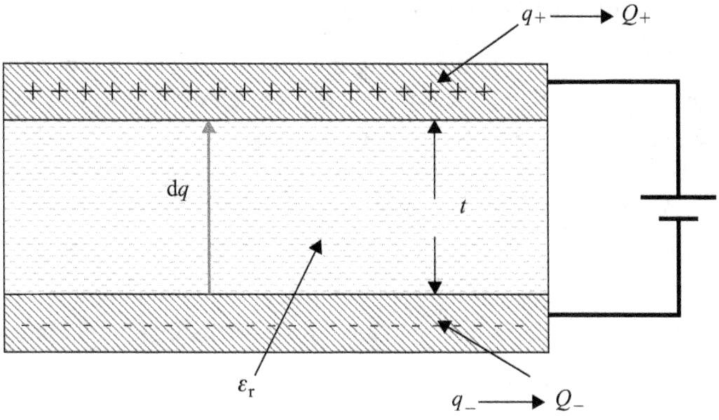

Figure 3.11 Work done in establishing charge

be directed normal to the plates as dictated by the boundary conditions represented by (2.21) and (2.23), and of constant magnitude. Hence

$$E = \frac{V}{t}$$

If the charge on the plates forms an infinitesimally thin surface charge layer (σ C/m^2) on the inner electrode surfaces, as it must do for a good conductor, then clearly for either plate ($Q = Q_+ = Q_-$)

$$Q = \sigma A \tag{3.18}$$

Therefore, energy stored

$$\begin{aligned} W_e &= \frac{1}{2}VQ = \frac{1}{2}(Ed)(\sigma A) \\ &= \frac{1}{2}E\sigma(Ad) \end{aligned} \tag{3.19}$$

Now if we apply Gauss' law to the field between the plates, we obtain

$$D = \sigma \quad (\text{C/m}^2) \tag{3.20}$$

$$\therefore\ W_e = \frac{1}{2}ED(Ad)$$

But Ad is the volume enclosed by the capacitor.

$$\therefore\ \text{Energy stored/unit volume} = \text{Energy density}$$

$$w_e = \frac{1}{2}ED \quad (\text{J/m}^3)$$

In vectorial terms this becomes

$$w_e = \frac{1}{2}\mathbf{D}\cdot\mathbf{E} \quad (\text{J/m}^3) \tag{3.21}$$

3.3.2 Force between plates of an air-spaced capacitor

Let us consider the problem of calculating the force required to separate the plates of an air-spaced parallel plate capacitor against the electric field due to voltage V (see Figure 3.12). We will assume that the charge Q is not altered as the plates are separated. (We could alternatively assume V to be held constant in which case Q would change and we would get a different result for force.) The initial stored energy is

$$U = \frac{1}{2}\frac{Q^2}{C}$$

The change in the stored energy ΔU brought about by the increased plate separation ΔZ is

$$\Delta U = \Delta\left(\frac{1}{2}\frac{Q^2}{C}\right)$$

But if Q remains constant then C must vary, in which case

$$\Delta U = \frac{1}{2}Q^2\left[\Delta\frac{1}{C}\right]$$

But $C = \frac{\varepsilon_0 A}{Z}$ (Farad) (for $\varepsilon_r = 1$) when a parallel plate capacitor is assumed.

$$\therefore\ \Delta U = \frac{1}{2}Q^2\Delta\left(\frac{Z}{\varepsilon_0 A}\right)$$

$$= \frac{Q^2}{2\varepsilon_0 A}[\Delta Z]$$

Now this supplied energy must be equal to the work done to separate the plate by a distance ΔZ. This work is given by the product: force $\times$ ΔZ.

$$\therefore\ \text{Force} = \left(\frac{1}{2}\frac{Q^2}{\varepsilon_0 A}\Delta Z\right)\Big/\Delta Z$$

$$= \frac{Q^2}{2\varepsilon_0 A} \quad \text{(Newton)} \tag{3.22}$$

Since the pressure on the electrode is $P_z A$ (N/m^2), and since $Q = DA$ from Gauss law, then we must have

$$P_z = \frac{D^2}{2\varepsilon_0} \quad (\text{N/m}^2) \tag{3.23}$$

Not surprisingly, given the symmetry of the Maxwell relations, this equation exhibits the same form as the equivalent expression for magnetic pressure.

3.3.3 Suspension criteria

Of course, although the above relationship (3.23) furnishes us with the basic equation, which underpins electrostatic levitation systems, we cannot ascribe the force acting on a capacitor electrode as being representative of electrostatic suspension, while it remains connected to an external circuit. To elevate a free float by electrostatic

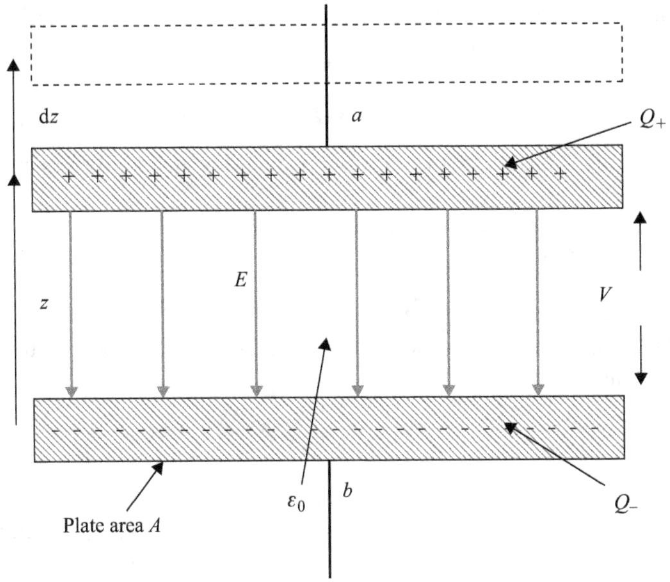

Figure 3.12 Electrostatic capacitor with movable electrode

forces, charge separation must occur within the conducting float, as suggested in Figure 3.13. Free electrons in the conducting float are drawn to the positive fixed electrode while positive charge is exposed to the negative electrode at the other end of the plate. If the capacitance between the float and the electrode pair is C, then the lifting force on the free plate is given by (3.22) but with a negative sign because upward movement in the $+z$ direction reduces C and hence U. In this case

$$F = \frac{q^2}{2\varepsilon_0 A} = \frac{q^2}{2\varepsilon_0 A}\frac{\Delta z}{\Delta z} = \frac{q^2}{2}\frac{\Delta \frac{1}{C}}{\Delta z} = -\frac{q^2}{2C^2}\frac{\Delta C}{\Delta z} \tag{3.24}$$

where A is the total electrode area $2\pi R^2$. Now, as we have seen, the ratio q/C for a simple parallel plate capacitor system is just the voltage V between the electrode

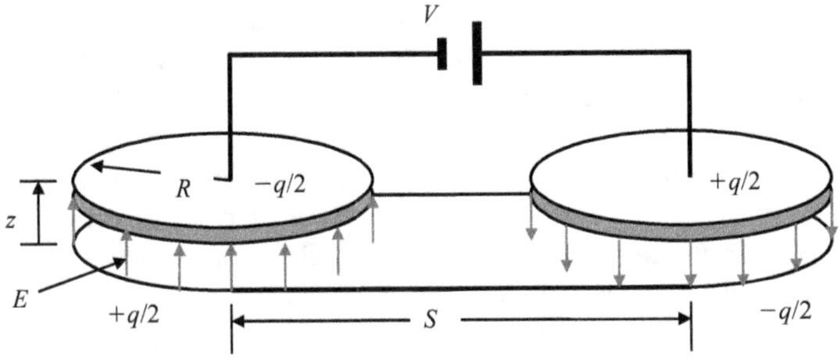

Figure 3.13 Schematic of levitation by electrostatic forces

plates and the float. Consequently, we can write (noting that the upward movement of the float implies the Δz is negative):

$$F_z = \frac{V^2}{2}\frac{\Delta C}{\Delta z} = \frac{1}{2}\frac{\varepsilon_0 A}{z^2}V^2 \quad \text{(Newtons)} \tag{3.25}$$

If the float exhibits a mass m, then equilibrium should be in equilibrium at a distance of separation $z = z_0$ given by

$$\frac{1}{2}\frac{\varepsilon_0 A}{z_0^2}V^2 - mg = 0 \tag{3.26}$$

where g is the gravitational acceleration. This equilibrium position is stable provided there is a force tending to restore the float to z_0 when it is displaced. For this to occur it is essential that

$$\left.\frac{\partial F_z}{\partial z}\right|_{z=z_0} < 0 \tag{3.27}$$

where $z = z_0$ denotes the equilibrium position. However, if V is constant, as it would be for a battery source, the application of the differential (3.27) to the force expression given in (3.25) yields

$$\left.\frac{\partial F_z}{\partial z}\right|_{z=z_0} = \frac{1}{2}\frac{\varepsilon_0 A}{z_0^3}V^2 \tag{3.28}$$

Clearly, the inequality in (3.27) is not satisfied, and the system is unstable for a DC applied voltage. The issue of stability in levitation systems will be examined in more detail in Chapter 4.

3.4 Lorentz effect

Eddy currents have been known to electrical engineers since the advent of inductors and transformers, especially those furnished with magnetic flux enhancing soft iron cores. These currents were initially merely unwanted sources of heat and inefficiency, but later, engineers began to convince themselves that it could actually be employed quite usefully. Ferrite-based electrical components get hot during operation due to the eddy currents in the core induced by the changing magnetic field – as explained below. Laminations are normally introduced to mitigate the effect. The faint hum in a mobile phone charger, for example which most phone users will be aware of, is caused by lamination movement, due to the same forces that provide levitation as we will see in section 3.4.2.

In addition to levitation, eddy currents are used gainfully in induction heating systems (Figure 3.14) where high temperature is created in the metal work-piece by immersing it in the time-changing magnetic field of an induction coil (Faraday effect). Other applications are in electromagnetic brakes and in induction cookers. In all cases the induced currents (eddy currents) result in Joule or I^2R heating in the conducting elements that are exposed to the time-varying magnetic field. In transformers and inductors, the energy loss due to these eddy currents in the ferromagnetic

Figure 3.14 Induction or Eddy current heating of a metal bar

core is in addition to the energy lost in the magnetisation/demagnetisation process, which as we have already seen, is proportional to the area enclosed by the material's hysteresis loop.

Levitation based on the generation of eddy currents is commonly referred to as electromagnetic levitation (EML). However, electromagnetic levitation is more correctly associated with electromagnetic pressure (see section 3.2.6). Consequently, here we prefer to refer to the kind of secondary or induced current levitation, outline in section 3.4.2, as eddy current levitation or ECL. The basic concept is illustrated schematically in Figure 3.3d and e.

3.4.1 Induced current

Electrical induction is usually illustrated in engineering texts by means of a schematic diagram, such as the one presented in Figure 3.15a. It purports to represent a C-shaped iron core wrapped with an N-turn winding carrying a current of *i* Amperes. For a steady current *i* and no electric field, the magnetic field generated in the core is given by Ampere's law (2.29) that yields

$$2\pi R H_{\mathrm{core}} = Ni$$

that is

$$B_{\mathrm{core}} = \frac{Ni}{2\pi \mu_r \mu_0 R} \quad \text{(Tesla)} \tag{3.29}$$

where R is the mean radius of the core. If the gap is small, such that field fringing is negligible, Maxwell's second law dictates that $B_g = B_{\mathrm{core}}$. Hence, we can write:

$$B_g = \frac{Ni}{2\pi \mu_r \mu_0 R} \quad \text{(Tesla)} \tag{3.30}$$

Now let us consider what happens if somehow a stationary charge q, indicated by the cross, were to be inserted into the gap. Since it is stationary, the Lorentz law (2.25)

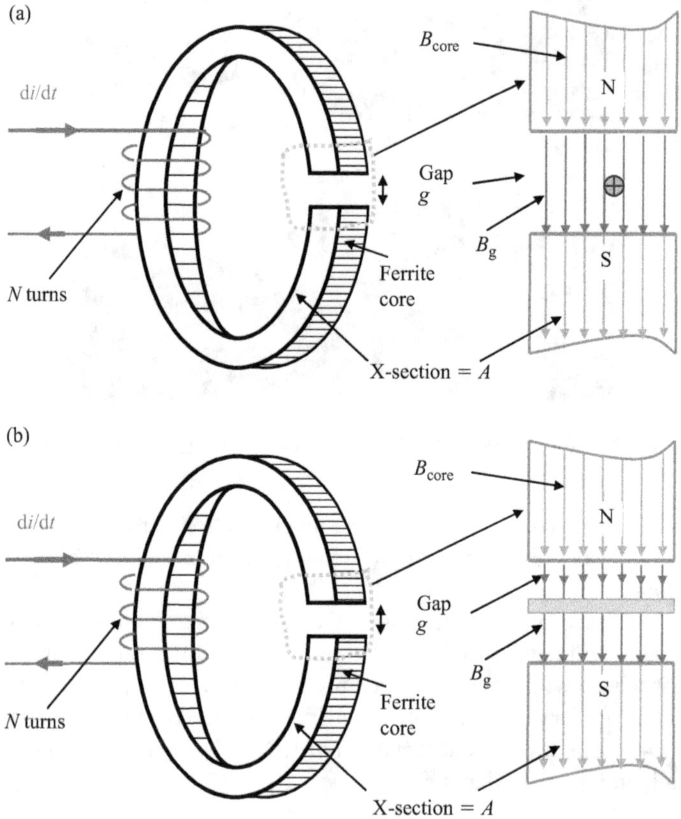

Figure 3.15 C-shaped soft iron core with N-turn winding and air gap: (a) with free charge in gap, (b) including conducting plate

reduces to $F = qE$. In this purely magnetic set-up there can be no E, so we must conclude that there is no force acting on the charge. This purely magnetic circuit, would seem to be unchanged if the current i is now an instantaneous value for a time-changing input current. The above equations are essentially unaltered except that the steady current i is replaced by an instantaneous current i, and one might be inclined to conclude that the force on a stationary charge in the air gap would still be zero. But this would be profoundly wrong. In fact, the free charge would be observed to move and the motive force can only come from the Lorentz first term, $F = qE$, since the initial charge velocity v, is still zero. An electric field exists in the gap. This field in the gap, which arises when i is time varying, is an induced electric field. It is of course calculable using Faraday's law or the fourth Maxwell equation (2.16). This phenomenon can easily be demonstrated by inserting a thin conducting plate in the gap of a soft iron inductor core such that it bisects the gap (see Figure 3.15b). Application of an AC current to the coil will result in free charge in the plate moving in circular paths within the metal sheet linking the magnetic field lines. These *eddy currents*, as they have been termed, will manifest themselves by gradual warming of the plate.

3.4.2 *Eddy current force*

More importantly, from a levitation perspective, eddy currents can also be generated in a conducting plate immersed in a magnetic flux, as shown in Figure 3.15b, even if the current in the coil and hence the flux is held constant. In this case, the 'change' required to induce the currents must come from plate movement, and the consequential Lorentz forces acting on the free charge in the plate. The situation is magnified in Figure 3.16.

Consider first movement of the plate to the right with velocity v_t. Free charge (q say) in the conducting plate, essentially electrons unattached to the fixed metallic ions, thus moves to the right with velocity v_t that being orthogonal to B_g, results in the Lorentz force $F_z = qv_t \times B_g$ acting in the z-direction. The charge is thus impelled to move with a velocity v_z so generating a further Lorentz force $F_t = qv_z \times B_g$ acting in the $-y$ direction. In other words, movement of the conducting plate to the right sets up an instantaneous restoring force, which resists any change being made, in accordance with Lenz's law. This law dictates that change in physical systems is always resisted (e.g. inertia in mechanical systems).

For movement of the plate in the x-direction with velocity v_n, it is less clear where the restoring force, if any, comes from since the vectors v_n and B_g are aligned with each other. If a Lorentz force exists it can only be associated with fringing fields B_y acting in the y and $-y$ directions. At the plate periphery therefore, where it penetrates the fringing field in the gap, a z-directed force $F_z = qv_x \times B_y$ is formed. This force produces a secondary z-directed current in the plate that combines with

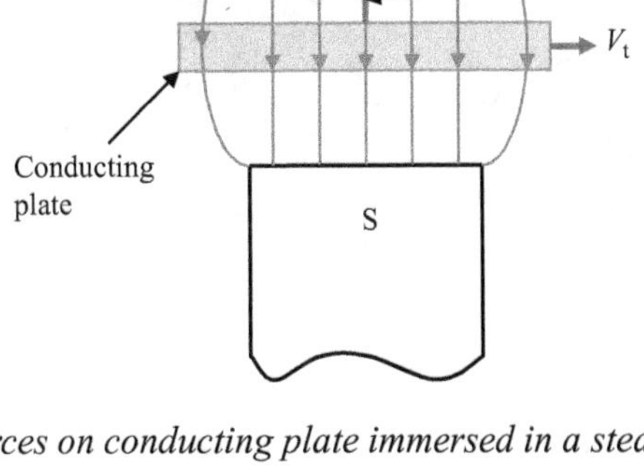

Figure 3.16 Forces on conducting plate immersed in a steady magnetic field

B_y to produce a downward restoring force $F_n = qv_z \times B_y$, again in accordance with Lenz's law.

The above forces are rather difficult to calculate for a conducting plate that exhibits finite conductivity (the issue is treated in Chapter 7), but fortunately it is easy if the intervening plate happens to be a perfect conductor or high temperature superconductor (HTS). In this case, for x-movement, a current is generated in the plate, which produces an opposing field that *totally* negates the original field in the gap. Consequently, the change in energy ΔW in the gap and hence the restoring force is given by

$$F_x \delta g = \Delta W = \frac{B_g^2}{2\mu_0} A \delta g \quad \text{(Newtons)}$$

where δg denotes a small movement of the float towards the upper pole. Thus, the pressure on an HTS float is

$$P_x = \frac{B_g^2}{2\mu_0} \quad (\text{N/m}^2) \tag{3.31}$$

Not surprisingly, this is no different to the expression developed earlier for the magnetic stress on the poles, in a purely magnetic levitation system (see (3.15)). In fact, the Maxwell stress within an electromagnetic field system is a quite general concept with wide application [18].

An example of levitation using eddy currents (ECL) is shown in Figure 3.17. A rapidly changing current in a set of coils produces a magnetic field that in turn produces an opposing current and magnetic field in the sample. The result is that the sample hovers inside the coils. The current can also be used to heat and ultimately melt the sample. The main reason for using ECL is that the sample can be held in either a neutral atmosphere or a vacuum and never comes into contact with a

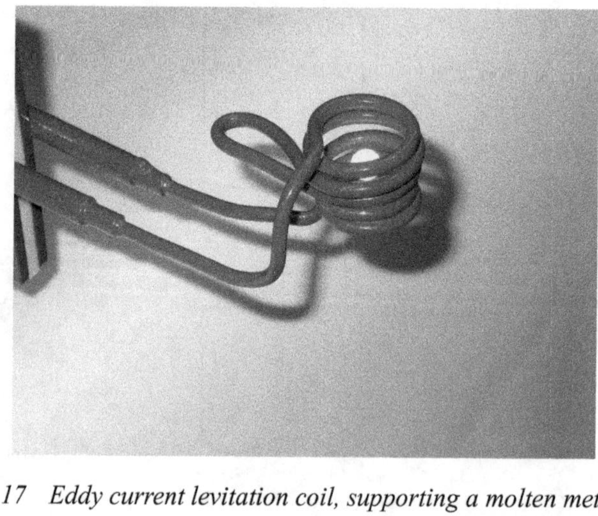

Figure 3.17 Eddy current levitation coil, supporting a molten metal sample

container wall. Thus, contamination can all but be eliminated. The illustration shows a single coil that is appropriate for a terrestrial laboratory where gravity provides the balancing force. In a space application, in microgravity, separate coils would be used for heating and levitation and the arrangement becomes symmetrical.

3.5 Levitation by electromagnetic pressure

Electromagnetic pressure is a phenomenon well known to science. Attempts to measure it were being pursued as early as 1901 by Ernest F. Nichols and Gordon F. Hull [19,20]. Their device, a more recent version of which is shown in Figure 3.18, consisted of a pair of small silvered glass mirrors suspended in the manner of a torsion balance by a fine quartz fibre within an evacuated enclosure. The torsion head to which the fibre was attached could be turned remotely to expose different mirrors to a light source. A beam of light was directed first on one mirror and then on the other, and the opposite deflections observed with mirror and scale. With this apparatus the experimenters were able to obtain an agreement between observed and computed radiation pressures within about 0.6%. The original apparatus is at the Smithsonian Institution in Washington, DC.

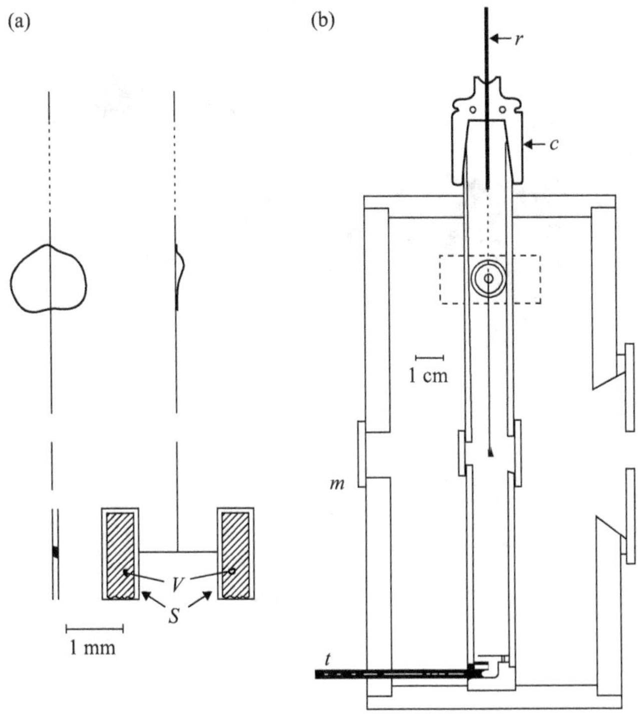

Figure 3.18 Nichols radiometer: (a) suspended system showing relative positions of vanes (V) and shields (S), and (b) radiometer case

The radiometer associated with Nicols and Hull is sometimes confused with a device invented by Sir William Crookes in 1873, an example of which is depicted in Figure 3.19. It has recently become an ubiquitous novelty ornament. While it is referred to as the Crookes radiometer, this is really a misnomer. When a radiant energy source is directed at a Crookes device, it becomes a heat engine. The operation of a heat engine is based on a difference in temperature that generates a mechanical output. In this case, the black side of the vane (see Figure 3.19) becomes hotter than the other side, as radiant power from a light source warms the dark side by black-body absorption, faster than the silver or white side, which reflects the radiation. Provided the chamber contains some air or gas (unlike the Nicols metre that ideally requires a vacuum) the local air near the black surface of each vane is 'heated up'. This means that nearby molecules become more agitated, increasing in speed when they touch the dark side of the vane, whereas elsewhere in the chamber and near the reflective vane surfaces molecular velocities are unchanged. The agitated air molecules near the 'hot' dark surfaces generate additional pressure on these surfaces, which moves the vanes forward with the black surface moving away from the light source. The opposite is the case in a radiometer, which truly measures radiation pressure.

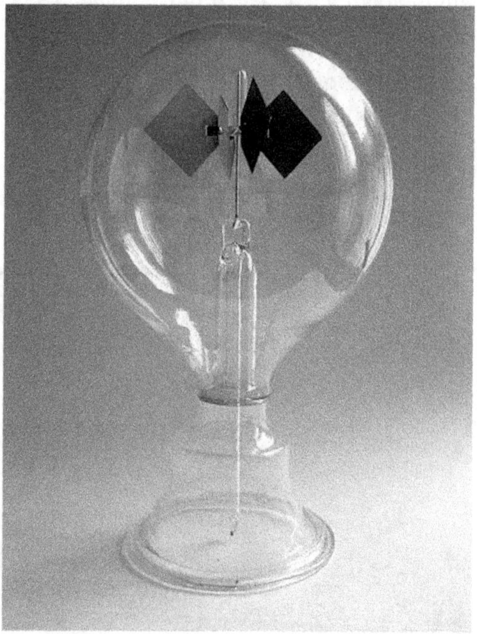

Figure 3.19 Crooke's radiometer

3.5.1 'Free space' waves

In Chapter 2, it has been demonstrated that the solution to the Maxwell equations in an infinite 'free' space is a plane electromagnetic wave (TEM wave) that can be expressed mathematically as

$$E_x = A \exp j(\omega t - \beta z) + B \exp j(\omega t + \beta z) \tag{3.32}$$

$$H_y = \frac{A}{\eta_0} \exp j(\omega t - \beta z) - \frac{B}{\eta_0} \exp j(\omega t + \beta z) \tag{3.33}$$

if z-directed, in a space defined by a Cartesian coordinate system. It is pertinent to note that the electric field E is normal to the magnetic field H, and both are normal to the direction of propagation. This is a general characteristic of TEM waves. E and H comprise two components; a forward or transmitted wave travelling in the $+z$ direction (first term) and a backward or reflected wave travelling in the $-z$ direction (second term). In air or vacuum each component travels at the speed of light (c). The properties of such a sinusoidally varying wave were considered in more detail in Chapter 2. However, it is useful to repeat here that the plane wave impedance η_0 is given by

$$\eta_0 = \sqrt{\frac{\mu_0}{\varepsilon_0}} \quad \text{(ohms)} \tag{3.34}$$

while the phase shift constant β is defined as

$$\beta = \frac{\omega}{c} = \omega\sqrt{\mu_0 \varepsilon_0} \quad \text{(rad/m)} \tag{3.35}$$

3.5.2 Pressure calculation

Now consider what happens if the entire x–y plane at $z = 0$ is covered by a flat perfectly conducting metallic sheet. At the surface of this sheet any tangential electric field if it were to exist would be immediately 'shorted out' by current flow in the sheet. The boundary condition on the electric field at the sheet (at $z = 0$) is that tangential E (i.e. E_x) must be everywhere zero for all time t. The only way this can be achieved is if $A = -B$. Mathematically, this implies that for $z > 0$:

$$\begin{aligned} E_x &= A[\exp j(\omega t - \beta z) - \exp j(\omega t + \beta z)] \\ &= 2A \exp j\omega t \sin \beta z \end{aligned} \tag{3.36}$$

This equation now infers that while E_x continues to vary sinusoidally at frequency ω as it must, it is no longer a propagating wave. The functionality with respect to z is independent of time, and E_x has the form of a 'standing wave', which exhibits fixed zeros at $\beta z = 0, \pi, 2\pi, \ldots, n\pi$ (n = even integer) and fixed maxima at $\beta z = \pi/2, 3\pi/2, \ldots, n\pi/2$ (n = odd integer).

Furthermore

$$H_y = \frac{2A}{\eta_0} \exp j\omega t \cos \beta z \tag{3.37}$$

The tangential magnetic field maximises at $z = 0$, in accordance with the field boundary condition, giving

$$|H_y| = \frac{2A}{\eta_0} \quad \text{(A/m)} \tag{3.38}$$

The presence of this tangential magnetic field at the surface of the perfectly conducting sheet implies, in accordance with Ampere's law, that a surface current $|J_{-x}| = |H_y|$ (A/m) flows in the sheet. This means that at the surface, charge is in motion in a direction normal to an adjacent magnetic field, in which case the Lorentz force law applies. The calculation of this force and the pressure on the plate P generally parallels the argument preceding (3.26) and, not surprisingly, leads to a similar result for pressure. Namely

$$P = \frac{B_y^2}{2\mu_0} \quad (\mathrm{N/m^2}) \tag{3.39}$$

In general, what this means is that when an electromagnetic wave impinges on a perfectly conducting surface, a pressure is exerted on the surface given by

$$P = \frac{B_t^2}{2\mu_0} \quad (\mathrm{N/m^2}) \tag{3.40}$$

where B_t is the tangential magnetic flux density at the surface. The Nicols radiometer attempts to measure this pressure, at optical frequencies, on highly conducting metallic vanes. Levitation based on this type of pressure is discussed in more detail in Chapter 9 (using Maxwell stress/momentum).

3.6 Summary

Electromagnetism is a mature and settled scientific topic, as we have seen in Chapter 2, and consequently it is hardly surprising that the electric and magnetic field distributions in devices designed to provide levitation or suspension are generally easily formulated and evaluated. Quite often, simple closed form equations are available, as we have seen here, to compute gap fields. If they are not then modern software packages based on finite difference or finite element techniques will accommodate the more complex geometries. Once the fields are known, as this chapter emphasises, the computation of suspension/levitation forces mainly entails an examination of incremental energy changes for small float movements.

Following a hopefully informative 'potted history' of the development of the laws and rules governing electromagnetic fields and levitation/suspension, the fundamental force equation that relates charge and electric and magnetic fields, namely the Lorentz law, is examined with reference to some practical manifestations. Magnetostatic or permanent magnet structures for levitation are introduced in section 3.2, where magnetic energy equations emanating from the Maxwell equations are developed and subsequently applied to non-linear materials. The basic forces and stresses that occur in permanent magnetic circuits, incorporating air gaps, as required by levitation mechanisms, are discussed in section 3.2.5. In a similar manner, the fundamental energy relations and suspension forces that occur in electrostatic devices comprising an electrode and a conducting float are treated in section 3.3.

Time-varying magnetic fields also produce levitation forces through the mechanism of induction and eddy currents as elucidated in section 3.4, while electromagnetic pressure, associated with high frequency trapped waves, is identified as a source of levitation in section 3.5.

3.7 References

[1] Ferrari, R. *An introduction to electromagnetic fields*. New York, NY: Van Nostrand Reinhold Co. Ltd.; 1975

[2] Johnk, C.T.A. *Engineering electromagnetic fields and waves*. New York, NY: John Wiley & Sons; 1988

[3] Feynman, R.P., Leighton, R.B., & Sands, M. *Lectures in physics*, vol. 2. London: Addison-Wesley Publishing Co.; 1972

[4] Morse, P.M., & Feshbach, H. *Methods of theoretical physics*. New York, NY: McGraw-Hill Book Co.; 1953

[5] Kraus, J.D. *Electromagnetics*. London: McGraw-Hill Book Co.; 1984

[6] Lorrain, P., & Corson, D. *Electromagnetic field and waves*. San Francisco, CA: W.H. Freeman & Co.; 1962

[7] Baden-Fuller, A.J. *Engineering electromagnetism*. New York, NY: John Wiley & Sons; 1993

[8] Hammond, P. *Applied electromagnetism*. Oxford: Pergamon Press Ltd.; 1971

[9] Braunbeck, W. 'Free suspension of objects by electric and magnetic fields'. *Zeitschrift für Physik*. 1939;**112**:752–763

[10] Beams, J. *Suspension of rotatable bodies*. US patent no. 2,256,937, 1941

[11] Beams, J. 'Production and use of high centrifugal fields'. *Science*. 1954; **120**:619–625

[12] Bachelet, E. *Levitating transmitting apparatus*. US patent no. 1,020,942, 1912

[13] Kemper, H. *Monorail vehicle with no wheels attached*. German patent no. 643,316, 1934

[14] Livingston, J.D. *Rising force*. Cambridge, MA: Harvard University Press; 2011

[15] Peskin, M.E., & Schroeder, D.V. *An introduction to quantum field theory*. Boulder, Colorado: Westview Press; 1995

[16] Silberberg, M.S. *Chemistry: the molecular nature of matter and change*. 2nd edn. Boston, MA: McGraw-Hill; 2000

[17] Moon, F.C. *Superconducting levitation*. New York, NY: John Wiley & Sons Inc.; 1994

[18] Carter, G.W. *The electromagnetic field*. London: Longman Group Ltd.; 1967

[19] Nichols, E.F., & Hull, G.F. 'A preliminary communication on the pressure of heat and light radiation'. *Physical Review*. 1901;**13**:307–320

[20] Nichols, E.F., & Hull, G.F. 'The pressure due to radiation'. *The Astrophysical Journal*. 1903;**17**(5):315–351

Chapter 4
Magnetic levitation and suspension

Nothing is too wonderful to be true if it be consistent with the laws of nature.

Michael Faraday

4.1 Magnetic circuits

To overcome the Earnshaw instability hurdle, which was introduced in Chapter 3, magnetic levitation systems using DC magnetic circuits must incorporate an 'active' element. This can be done in several ways: through physical motion (e.g. spin), through eddy currents or through electronic control. The employment of eddy currents enhanced by diamagnetic or superconducting techniques to secure stable levitation will be examined in Chapter 7. Here we will focus on magnetic systems that achieve stability through electronic control.

The forces and the stresses, associated with the magnetic fields produced by magnetic poles in close proximity, were examined in Chapter 3 and the relevant equations for determining the interactions between adjacent poles are provided there. However, in practice, calculations of levitation forces are complicated by the fact that floats exist in an air gap and these interaction spaces form elements in a magnetic circuit. Consequently, the determination of the gap fields requires a prior analysis of flux relationships within the circuit as a whole. A schematic illustration of a possible embodiment of an elementary active magnetic levitation system is presented in Figure 4.1, both for a repulsion example (Figure 4.1a) and an attraction example (Figure 4.1b). The representation clearly shows that in either case the magnetic circuit includes a shaped float (permanent magnet), a shaped electromagnet (soft iron) and a pair of floatation gaps that are presumed to be monitored by an infra-red feedback system. Other sensing mechanisms based on capacitive, inductive or Hall effects, could also be used.

Today, magnetic cores such as those depicted in Figure 4.1 would normally be developed by resorting to sophisticated electromagnetic simulation software systems [1,2]. These elaborate comprehensive commercial packages permit accurate representation in three dimensions of the electromagnetic boundary value problem, including complex, non-linear properties of bulk materials. In the majority of packages, the model is then solved by either finite difference or finite element techniques. However, successful operation of such software requires an intelligent application and a good understanding, by the user, of the analyses techniques on which they are

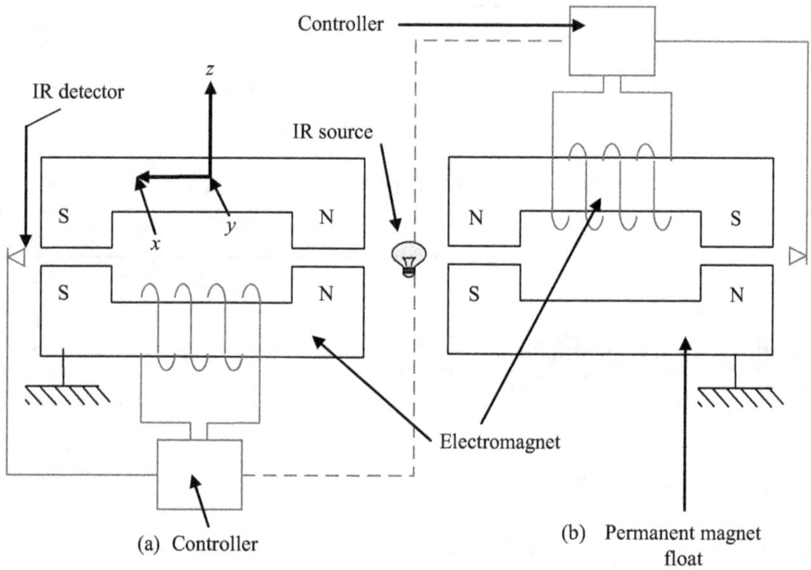

Figure 4.1 Basic magnetic circuits for active magnetic levitation: (a) repulsion; (b) attraction

based. In addition to Maxwell's equations a good knowledge of approximate computational methods is also an advantage in setting up plausible trial conditions and in checking the validity of the computed results. It is well known that without due care to model definition programmes of this kind can converge on solutions that are spurious or erroneous. Approximate solutions for magnetic circuits are introduced in section 4.1.1.

4.1.1 Electrical analogies

For electrical engineers familiar with electrical circuit theory and network theorems, probably the most accessible approach to the analysis of magnetic circuits is achieved by creating a plausible analogy with electrical circuit theory. By so doing 'tried and tested' methods become available to the design of magnetic structures. A simple magnetic circuit depicted alongside an analogous electrical construct is shown in Figure 4.2. The magnetic circuit comprises a current carrying winding ('battery') that drives a magnetic flux ('current') around a core that is predisposed to impede the flux ('resistance').

The relationships between the magnetic circuit parameters can be deduced by application of the third Maxwell equation (2.3). There are very few phenomena in electrical science for which the linkage to Maxwell's equations is not available. If the current I is non-varying then (2.3) reduces to:

$$\text{circ}\,\mathbf{H} = NI \tag{4.1}$$

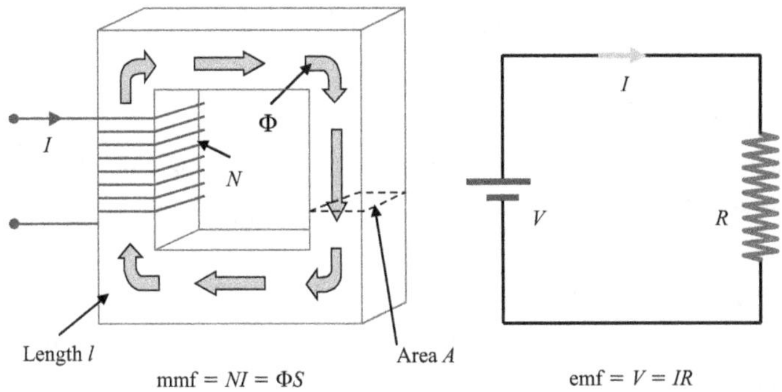

Figure 4.2 Equivalence between magnetic and electrical circuits

for an N turn coil. In the case of an elementary core of effective length l, as shown in Figure 4.2, the circulation integration of (4.1) simplifies to (see Appendix I):

$$NI = Hl \tag{4.2}$$

Here flux leakage is assumed to be negligible, which is usually an acceptable simplification. In circuit terms the coil is seen as the 'battery' of the magnetic arrangement exerting a magneto-motive force (mmf) equal to NI that generates magnetic flux (Φ) in the core. In the equivalent electrical circuit, the battery voltage or electro-motive force (emf) drives a current (I) around the loop that is analogous to the magnetic flux. It is pertinent to observe here that the electrical circuit is governed by the fourth Maxwell equation (2.4) for time invariant conditions.

The relationship between flux Φ, and the magnetising field H, depends on the nature of the material (see Chapter 2). It can be linear, in which case its relative permeability is a constant scalar quantity, or it can be non-linear in which case graphical representations of the relationship between flux and magnetising force must be employed as we will see below. For the linear case we can write:

$$\mathbf{B} = \mu_r \mu_0 \mathbf{H} \tag{4.3}$$

In the absence of magnetic field leakage, and for a core of regular cross-section within which the flux density vector **B** is normal to the cross-section and of constant magnitude, the flux calculation associated with Maxwell's second equation (2.2) simplifies to:

$$\Phi = BA = \mu_r \mu_0 HA \tag{4.4}$$

Hence using (4.2) we get:

$$\text{mmf} = NI = \frac{l}{\mu_r \mu_0 A}\Phi \tag{4.5}$$

By analogy with Ohms law in electrical circuit theory (emf $= RI$) (4.5) is usually written:

$$\text{mmf} = S\Phi \tag{4.6}$$

where S is referred to as the core reluctance and is given by

$$S = \frac{l}{\mu_r \mu_0 A} \quad (\text{Henry}^{-1}) \tag{4.7}$$

The electrical equivalent of S is provided by the resistor, namely R (Ohms). For a linear material it has a not dissimilar mathematical form, namely:

$$R = \frac{l}{\sigma A} \quad (\text{Ohms } (\Omega)) \tag{4.8}$$

Here l is the resistor length, A is its cross-sectional area and σ (mhos/m) is the conductivity of the resistor material.

4.1.2 Example calculation

Magnetic circuits [3,4] can manifest themselves in a variety of forms depending on the application, ranging from transformers, to rotating machines, to magnetic focusing systems to magnetic bearings, not to mention many more less well known uses of the technology. The typical problem illustrated in Figure 4.3a comprises a magnetic core of more than one branch with perhaps a current carrying winding on one or more branches and an air gap in another. For a core of known dimensions and for a predefined ampere-turns level in the winding on one of the limbs of the core, the usual question is: what is the gap flux density? Or for a levitation system perhaps, where the gap flux needed for levitation is known, what level of ampere-turns is required to achieve this with the given core?

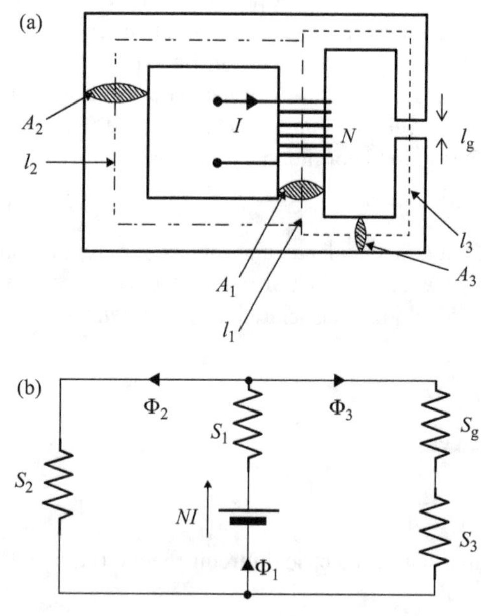

Figure 4.3 (a and b) Magnetic circuit example with winding and gap

For problems of this type based on the use of high permeability cores it is possible, with minimal error, to presume that the leakage magnetic flux around the core is negligible. Consequently, in Figure 4.3a, with NI defined, the unknowns to be determined are the fluxes in the three limbs of the magnet – namely Φ_1 in limb 1 with cross-sectional area A_1 and effective length l_1, Φ_2 in limb 2 and $\Phi_3 = \Phi_g$ in limb 3. To solve for these three unknowns we need three independent equations relating one to the other. These equations are formulated by setting up an analogous electrical circuit as shown in Figure 4.3b. Here the reluctances of each limb S_1, S_2, S_3 and S_g for the gap, are represented by 'resistors', while the battery voltage is replaced by winding mmf and the currents in the branches by fluxes. Circuits of this nature are amenable to Kirchhoff's voltage and current laws, and application of these elicits the following simultaneous equations:

$$NI = \Phi_1 S_1 + \Phi_2 S_2 \qquad (4.9)$$

$$NI = \Phi_1 S_1 + \Phi_3 (S_3 + S_g) \qquad (4.10)$$

$$\Phi_1 = \Phi_2 + \Phi_3 \qquad (4.11)$$

where for the given core:

$$S_n = \frac{l_n}{\mu_r \mu_0 A_n} \quad (n = 1, 2, 3) \qquad (4.12)$$

and

$$S_g = \frac{l_g}{\mu_0 A_3} \qquad (4.13)$$

Hence, the solution of (4.9), (4.10) and (4.11) for known values of NI and S_1, S_2, S_3 and S_g leads to desired values for Φ_1, Φ_2 and $\Phi_3 = \Phi_g$.

4.1.3 Inclusion of permanent magnet

For magnetic circuits formed from magnetic components that display linear characteristics, section 4.1.2 provides the essential elements of an approximate but reliable method of assessing flux relationships. Unfortunately, in the most useful and interesting applications of magnetic materials, alloys are frequently used that are often anything but linear in magnetisation terms. Permanent magnets formed from these materials can no longer be ascribed a representative fixed permeability value, but rather manufacturers have to supply users with measured tabular or graphical details which present the flux, or the flux density, in a well-defined sample of the material, as a function of mmf. When such materials are used in a magnetic circuit it becomes necessary to accommodate the graphical information into the analysis procedure. Figure 4.4 shows how this is done for a simple permanent magnet (C-magnet).

The sketched hysteresis loop (light unbroken trace) in Figure 4.4b is intended to be representative of a manufacturer's measured curve for a material such as a ferrite. It is clearly very non-linear, with B_c changing quite rapidly as the mmf (H_c) is raised or lowered in some parts of the loop and hardly varies at all in the 'saturated' portions. In terms of an analysis of the loop fluxes, it represents one equation in the

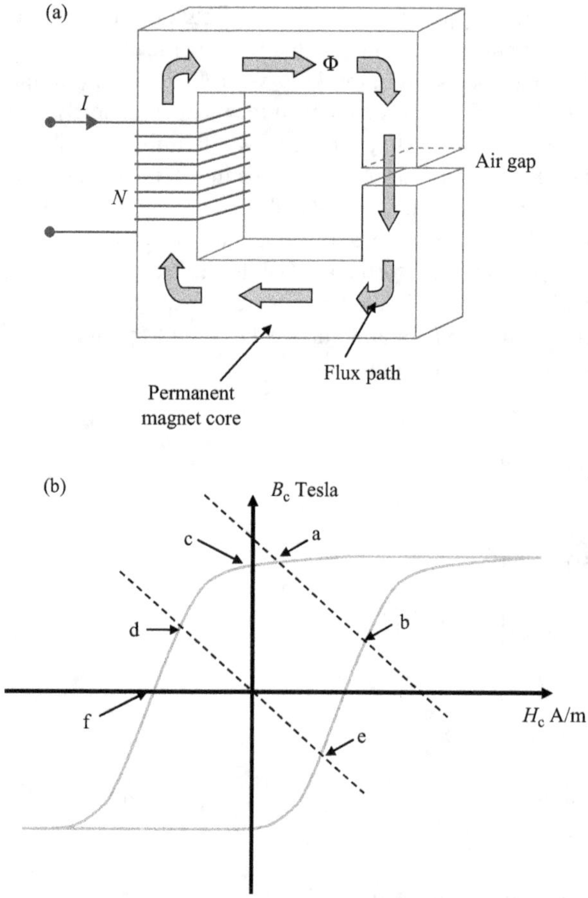

Figure 4.4 Permanent magnet core with winding and air gap (a) core construction, (b) core hysteresis loop

computational process. A solution for the circuit in Figure 4.4a obviously requires a graphical determination of where on the loop the magnet is 'resting' under the influence of the winding current and in the presence of the gap. To determine this point, we need at least one more equation that can be plotted graphically on the same B_c–H_c axes.

The required equation is obtained by applying Ampere's law (2.3) to a circulation path around the core (see Chapter 2). As in the previous section, leakage in circuits of this description can safely be presumed to be negligible, in which case (2.3) becomes:

$$H_c l_c + H_g l_g = NI \tag{4.14}$$

where l_c is the effective length of the core and l_g is the gap length. At the gap the second Maxwell equation applies, which in the absence of leakage yields:

$$B_g = B_c \tag{4.15}$$

Hence given that air has an absolute permeability μ_0, which is a constant, we can write:

$$H_g = \frac{B_g}{\mu_0} = \frac{B_c}{\mu_0} \tag{4.16}$$

Substitution of relationship (4.16) into (4.14) generates:

$$B_c = -\frac{\mu_0 l_c}{l_g} H_c + \frac{\mu_0 NI}{l_g} \tag{4.17}$$

This is the equation of a straight line on the B_c–H_c axes with a negative slope equal to $-\mu_0 l_c / l_g$ and intercept $\mu_0 NI / l_g$ on the B_c axis. Possible representations are depicted on the hysteresis graph in Figure 4.4b.

The upper dashed line that intercepts the hysteresis loop at points (a) and (b) is representative of a geometry comprising a moderate air gap with a finite current in the winding, while the lower dashed curve with intercepts at (d) and (e) retains the gap but has zero winding current. Evidently, there are two solutions in each case, but in practice only one is relevant since the flux direction in the core is in one direction – either clockwise or anticlockwise. Assuming that this known direction is represented by the upper branch of the hysteresis loop then solutions (a) and (d) are appropriate. In order to produce a core (and hence gap) flux density close to the saturation level at point (a) the winding current is high and represented by point (g) where the negatively sloping line intercepts the H_c axis. From (4.17), with B_c set to zero, the required ampere-turns can be deduced.

Changes in the winding current move the blue dashed line to the right (rising current) or left (reducing current) with point (a) moving closer to saturation or towards the slope on the left of the hysteresis loop. When the current drops to zero the dashed line passes through the origin and intercept the hysteresis loop at point (d). The flux in the core represented by this point is termed the 'remanent flux' in the presence of the given gap. If the gap is zero, the slope of the intercepting line becomes infinite (vertical line) and remanence occurs where the hysteresis curve crosses the B_c axis. In electrical terms the air gap acts as a high impedance in the circuit and reduces the flux (current) in the circuit to a level, represented by point (d), which is well below the zero gap case (point (c)). In the limit when the line passes through point (f), the magnetic flux density in the core is theoretically zero – core demagnetized. This condition requires a negative magnetising force ($-H_c$) dictated by the distance between point (f) and the origin. The force is termed the coercive force for the core material, and the larger is this force the more difficult it is to demagnetize the magnet. This property is desirable in many magnetic systems, including levitation systems, and to satisfy the demand, magnets that incorporate exotic rare earths such as Alnico (aluminium–nickel–cobalt) Samarium–cobalt and Neodymium (iron–neodymium–boron) have evolved in order to procure cores exhibiting very high coercive force levels [5].

There are some important and interesting differences between these permanent magnet materials. While Alnico, and ferrite magnets, can be fully demagnetized by applying the appropriate coercive force, with little flux density recovery when the force is removed, this is not true for the rare earth magnets [6]. With samarium–cobalt for example, while it is obviously still possible to eliminate all magnetic flux from the core, by applying a demagnetizing current at the level of the coercive force, as soon as this force is removed the flux density 'springs back' to almost its previous remanence level. Demagnetisation (long term) does not really occur, at least at room temperature, and the coercive force is termed 'intrinsic' to imply that it is the force required to procure very temporary negation of the intrinsic flux density. For samarium–cobalt the intrinsic coercive force can be from 20 to 50 times that of Alnico.

Needless to say, if one moves beyond the elementary circuit depicted in Figure 4.4a, to circuits with, perhaps, multiple sections comprising different alloys, with differing hysteresis characteristics, graphical procedures quickly become impractical. In this event there is little alternative but to resort to a sophisticated electromagnetic solver [1,2] and trust to the availability of sufficient computing power. In such software packages, hysteresis curves, for a range of relevant alloys, can be stored in tabular form. The analysis programme will repeatedly access these tables, as and when it needs to, as it iterates towards a unique flux solution for the network of magnets and gaps.

4.2 Levitation dynamics

A major and obvious difference between the magnetic circuits discussed earlier, and levitation circuits, is the presence of a section that 'floats'. Crucially, this implies continuously varying widths for the air gap(s). As we have already seen, an air gap in a magnetic circuit presents a very high reluctance to the flux in the core and this reluctance is very sensitive to the gap width (4.13). This has major significance for the maintenance of stability of the levitation, or suspension, set up by electronic means.

The theoretical development outlined in section (4.2.1) is by no means comprehensive, and is merely provided to illustrate the steps required to accommodate a moving element in a magnetic circuit. In practice, a real technical system involves stabilisation for several degrees of freedom, which entails the incorporation of a range of sensors, actuators and electronic control circuits. More complete treatments of more realistic scenarios can be found in references 6–10. Here we examine linear control using analogue techniques for an elementary suspension system as illustrated in Figure 4.5.

As has already been suggested by Figure 4.1, the float within a magnetic system can either be in repulsion mode (levitation – Figure 4.1a), or attraction mode (suspension – Figure 4.1b). In the primary direction of motion of the float (usually denoted as the z-direction), the former is the more stable arrangement. This is easily explained by reference to Figure 4.6, which presents a typical force profile as a function of separation distance for an adjacent pair of U-shaped magnets. Not unexpectedly the force diminishes rapidly with gap width (line with triangular symbols). It can apply to magnets with poles aligned for repulsion or for attraction. For levitation, the float would normally exhibit a weight ($m_f g$, g being gravitational acceleration) that lies between the maximum lift force when the gap is close to zero,

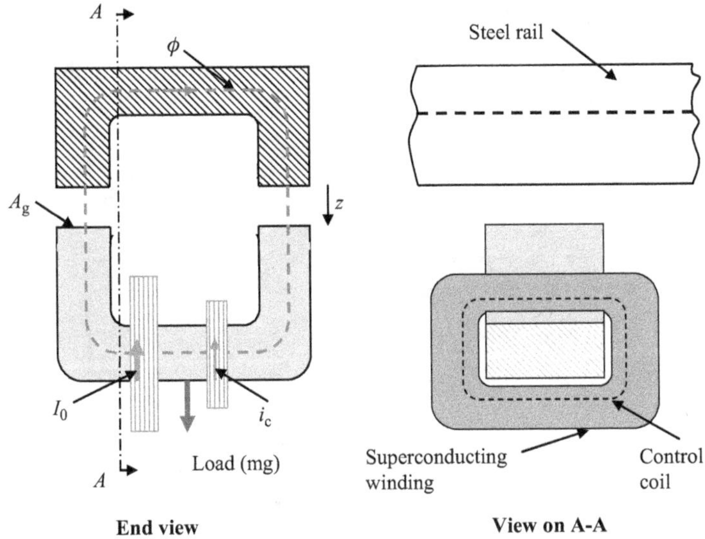

Figure 4.5 Schematic of feedback controlled magnetic float below iron rail

and a force representative of a distance beyond the desired location for stable levitation, denoted by the light horizontal line at level 5 in the diagram. The point of interception between the line and the curve represents the equilibrium condition for the float, where the magnetic force balances the gravitational force. In the levitation scenario (Figure 4.1a) with the float above the stator, on one hand, movement of the float downwards (decreasing gap) below the equilibrium position results in a restoring force located in the portion of Figure 4.6 where the green curve is above the red line. On the other hand, upwards movement away from equilibrium (increasing gap width) results in the gravitational force becoming predominant and again restoration to equilibrium is enforced. So for z-directed movement the magnetic levitation system is stable. This is not the case for suspension (Figure 4.1b). When the float moves upwards away from the equilibrium position (decreasing gap), the increasingly predominant magnetic force now encourages this movement with the magnets eventually clamping together (zero gap). Movement downwards from equilibrium again means that the gravitational force over-rides the magnetic force, but now in the unstable direction such that the float falls away. Magnetic suspension is thus intrinsically unstable for vertical motion.

This difference in behaviour between a levitated and a suspended magnet in a magnetic circuit can be explained by examining energy storage and 'the principle of least action'. In any magnetic circuit, however formed, the magnetic flux will distribute itself around the system so that the quantity of stored magnetic energy is a minimum. For a circuit with a variable gap, as in levitation or suspension systems, the minimum occurs in the 'attractive mode' for zero gap width. It requires work to open up the gap against the magnetic force of attraction, so any movement away

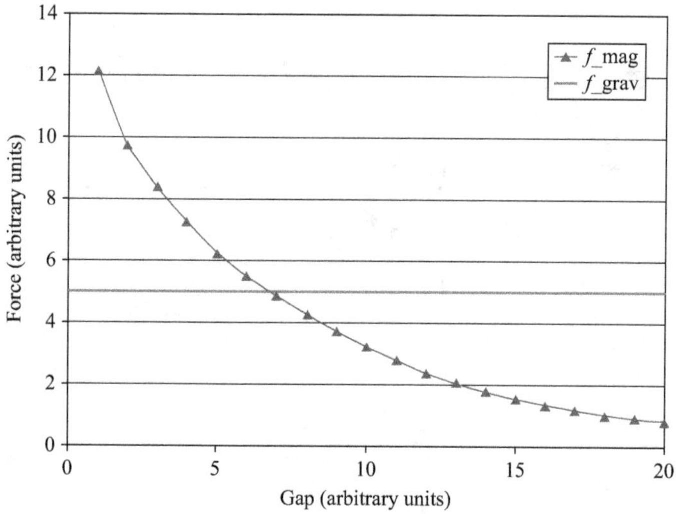

Figure 4.6 Levitation/suspension force for a magnetised float within a magnetic circuit

from the zero gap condition involves increasing magnetic energy storage. This means that magnets in repulsion store more and more energy as the gap between like poles is forced towards zero resulting in a storage maximum at this point. Hence movement away from the contact situation in any direction entails loss of stored magnetic energy – in gravitational terms moving downhill. But the float (Figure 4.1a) as it rises gains potential energy, so in total energy terms the z-moving float transitions from high magnetic energy through a minimum to high potential energy. At the minimum stable equilibrium exists for z-movement. Unfortunately, for lateral movement gravitational energy remains constant that means the total energy continues to sink (Figure 4.7). At the 'stability point' the float is on a 'saddle' or a ridge between energy peaks that it can easily 'slide off' as the system seeks a lower energy arrangement. As anyone, who has worked with magnets in repulsion, knows it is impossible to maintain equilibrium without a lateral guiding mechanism (see Figure 3.3a).

For magnets in suspension, as in Figure 4.1b, it has already been noted that minimum energy occurs where the magnets are in contact – zero gap. Work has to be done to increase the gap, thus producing increased stored magnetic energy. The rate of increase of this energy initially overwhelms the slowly diminishing gravitational energy stored in the float. So the total energy stored rises. But beyond equilibrium where the magnetic force just balances the gravitational force, the magnetic field disappears rapidly and gravitational effects take over with the total energy on a slow downwards slope. Consequently, at equilibrium, the float position represents a total energy peak, which is a point of instability. To maintain stability at the equilibrium position, sensors and feedback control, as suggested in Figure 4.1b, are essential. However, flotation by suspension wins over levitation insofar as it is intrinsically stable for lateral displacement. Apart from electromagnet current adjustments to

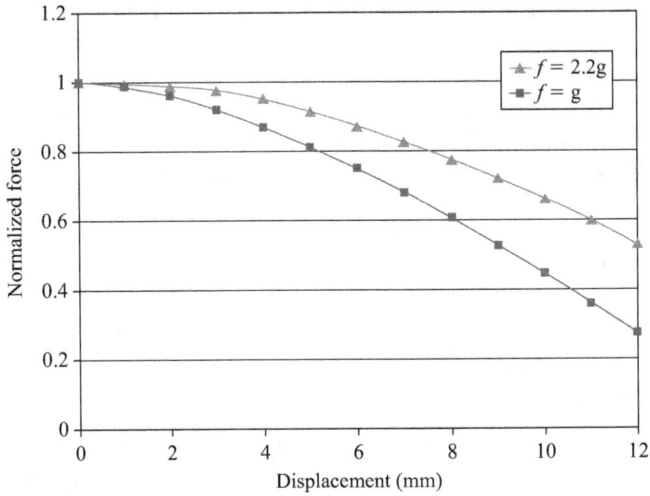

Figure 4.7 Lift force as a function of lateral displacement for levitated U-shaped magnet

compensate for the diminishing magnitude of the lift force (as suggested by Figure 3.8b for two different float weights) that the control system is designed to provide, the float experiences a restoring force, as shown in Figure 4.8b. This force draws the float back into alignment with the stator.

4.2.1 Dynamic equations

The stability of a float in a magnetic levitation or suspension system, which has been described in general terms in the preceding section, can of course be examined in detail using the language of mathematics. Figures 4.6–4.8 have been generated from a simple mathematical model that we will construct in this section. Models of this description, which enhance our understanding of the relationships between force, motion and energy of a magnetically floating body, usually adopt the assumption that the feedback arrangements are linear. Float motion is governed by Newton's laws while the magnetic circuit is modelled using Maxwell's equations (see section 4.1). For one degree of freedom suspension system, as depicted in Figure 4.5, where vertical movement is directed along the z-axis, application of Newton's laws and magnetic circuit relations leads to the dynamic force equation, namely:

$$m_f \frac{d^2 z}{dt^2} = m_f g - F \tag{4.18}$$

where F is the instantaneous levitation force, m_f is the mass of the float and g is the gravitational acceleration. The electrical circuit behaviour leads to the relation [5]:

$$N \frac{d\Phi}{dt} = -IR + V_0 + V_c \tag{4.19}$$

where N is the primary coil turns and R is the primary coil resistance, V_0 is the voltage across the main coil, while V_c is the voltage applied to the much smaller control loop that has negligible resistance. The DC voltage V_0 provides the steady magnetising current I_0 in the primary coil while V_c developed by the feedback circuit generates a control current to maintain the suspended mass with a requisite degree of stiffness at the equilibrium position. In linear control theory the appropriate step to take

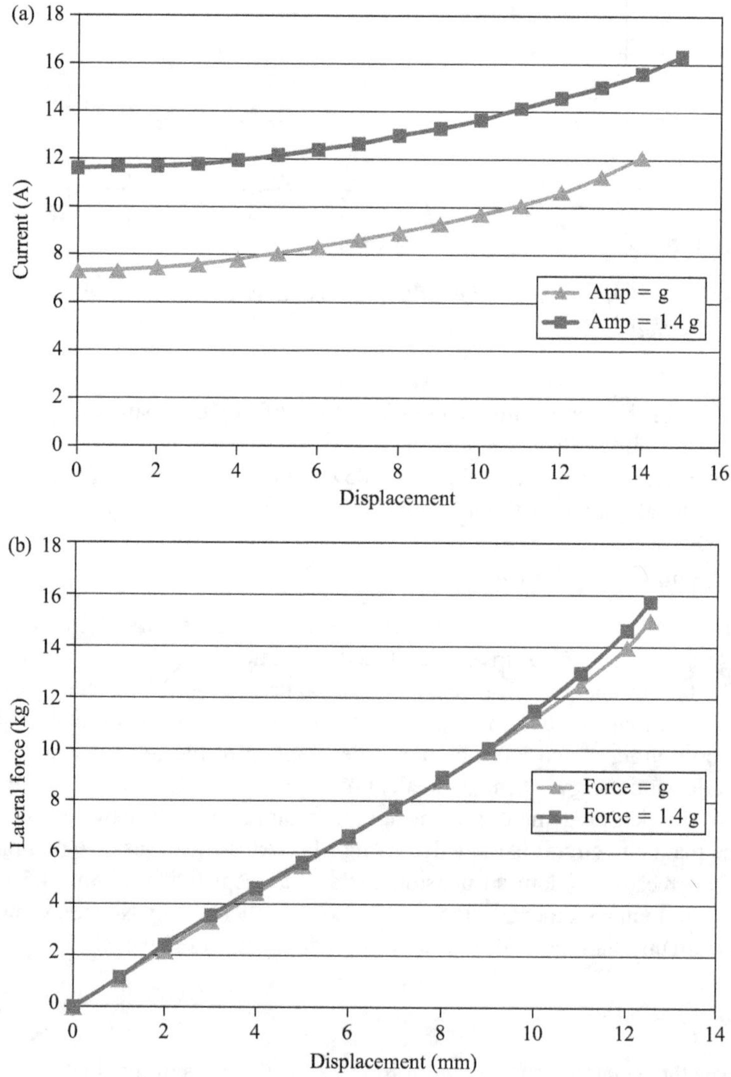

Figure 4.8 Lateral restoring force on suspended U-shaped magnet as a function of displacement. (a) Variation of electromagnet current to maintain lift, (b) restoring force as a function of lateral displacement at constant lift

at this juncture is to express V_c as a linear combination of the state variables, namely float position, instantaneous velocity and acceleration (flux). Hence, we choose to form the following relationship:

$$V_c = K_1(z - z_0) + K_2\frac{dz}{dt} + K_3\frac{d^2z}{dt^2} \tag{4.20}$$

where z_0 is the equilibrium position of the float and K_1, K_2 and K_3 are constants [6].

Now, magnetic circuit relations such as (4.6) dictate that:

$$\Phi = \frac{NI}{S} \tag{4.21}$$

where S is the total circuit reluctance. Hence, the magnetic force F in (4.18) is readily determined from the Maxwell stress (3.15) and has the form (for two gaps):

$$F = \frac{\Phi^2}{\mu_0 A_g} \tag{4.22}$$

where A_g is the gap cross-sectional area as shown in Figure 4.5.

The principal aim of any magnetic suspension system design is to determine values for the constants K_1, K_2 and K_3, which ensures that the system is dynamically stable for small disturbances from the equilibrium position of the float. Mathematically, this requires that perturbations are introduced as follows:

$$\begin{aligned} z &= z_0 + \zeta(t) \\ I &= I_0 + i(t) \\ S &= S_0 + S_g(t) \end{aligned} \tag{4.23}$$

Here $\zeta(t)$ is the float displacement from equilibrium (z_0), i is the control coil current that changes in sympathy with ζ. S_0 is the magnetic circuit reluctance at equilibrium, while S_g is the variation in reluctance due to changes in the gap width – that is

$$S_g = \frac{2\zeta}{\mu_0 A_g} = \alpha\zeta \tag{4.24}$$

We also know that at equilibrium:

$$V_0 = I_0 R \tag{4.25}$$

and that $m_f g = F$. Hence, from (4.22) we obtain:

$$m_f g = \frac{(NI_0)^2}{\mu_0 A_g S_0^2} \tag{4.26}$$

It is legitimate to presume that, for an effective dynamic control system, perturbations ζ will be small, in which case we can simplify the non-linear expressions for flux (4.21) and force (4.26) as follows:

$$\Phi = \frac{NI_0}{S_0}\left(1 + \frac{i}{I_0} - \frac{\alpha}{S_0}\zeta\right) \tag{4.27}$$

$$F = m_f g\left(1 + \frac{2i}{I_0} - \frac{2\alpha}{S_0}\zeta\right) \tag{4.28}$$

On combining (4.27) and (4.28) with (4.18) and (4.19), a pair of coupled linear differential equations result. These can be solved by employing Laplace transforms or matrix inversion techniques. The equations are:

$$\frac{\mathrm{d}^2\zeta}{\mathrm{d}t^2} - \frac{2g\alpha}{S_0}\zeta = -\frac{2gi}{I_0} \tag{4.29}$$

$$\frac{N^2}{S_0}\frac{\mathrm{d}i}{\mathrm{d}t} = -Ri + \frac{N^2 I_0 \alpha}{S_0^2}\frac{\mathrm{d}\zeta}{\mathrm{d}t} + K_1\zeta + K_2\frac{\mathrm{d}\zeta}{\mathrm{d}t} + K_3\frac{\mathrm{d}^2\zeta}{\mathrm{d}t^2} \tag{4.30}$$

The solution to these equations reveals that the magnetically suspended float with feedback control behaves not unlike a weight supported by a spring incorporating damping. This similarity of behaviour leads to the concept of 'magnetic stiffness'. In elastic systems of a mechanical nature the change in force with position is usually proportional to the displacement. The constant of proportionality, which relates the restoring force increase to the incremental displacement, is termed the elastic stiffness. By analogy, the magnetic stiffness is defined as the increment in the magnetic force acting on a float in response to a unit change in its position. Thus, mathematically, the magnetic stiffness κ for the system outlined above can be expressed as

$$\kappa = -\left.\frac{\partial F}{\partial z}\right|_{\mathrm{eq}} = \frac{2m_{\mathrm{f}}g\alpha}{S_0} \tag{4.31}$$

The stability of a magnetic suspension system employing feedback control essentially requires the control stiffness (or responsiveness) to exceed the magnetic stiffness, otherwise oscillation occurs. In terms of the mechanical spring analogy, oscillation equates to insufficient damping. The solution to the coupled equations (4.29) and (4.30) dictates that a properly damped performance requires that the following mathematical inequality should be satisfied:

$$K_1 > \frac{V_0\alpha}{S_0} \tag{4.32}$$

This equation points, not unexpectedly, to a magnetic circuit arrangement that displays low primary voltage to ensure effective control, and high reluctance to moderate the magnetic stiffness (4.31).

4.2.2 Elements of stability control

Once the magnetic circuit requirements and the desired dynamic conditions for a given suspension system have been determined, it remains to devise a control system capable of providing the requisite sensing, monitoring, processing and adjustment necessary for stable flotation. These electrical technologies reside in the disciplines of control theory and electronic circuit design and as such they are comprehensively treated in undergraduate texts [11–13]. Here we will merely touch upon the essential elements needed to affect a control solution for a magnetic suspension arrangement.

From our earlier discussion, it is clear that stable suspension entails regulating the current in an electromagnet in response to signals from suitable sensors strategically placed to monitor the flotation gap, as suggested in Figure 4.9. Control of the electromagnet current alters the lift force versus air gap characteristics as shown

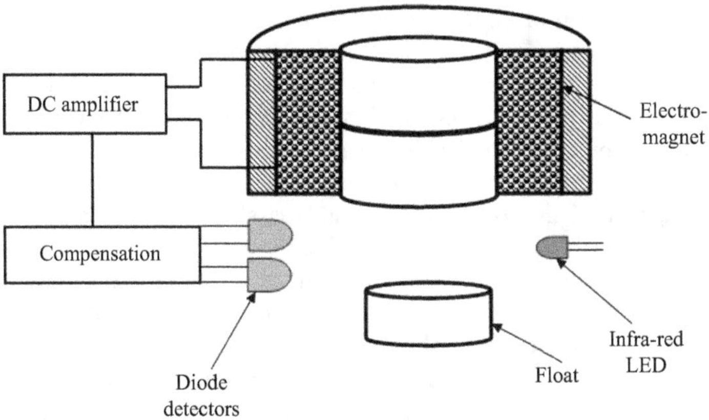

Figure 4.9 Active suspension of an iron float using infra-red sensing

in Figure 4.10, with the open loop force rising with current at a fixed gap. When feedback is introduced (closed loop), the current in the electromagnet is made to rise when the gap increases, as detected by the sensor, and fall when the gap decreases, so that the lift force now behaves in accordance with the dotted curve in Figure 4.10. With photocell sensing of the gap width, as suggested in Figure 4.9, control depends on the amount of light reaching the cell. If the float, in this example a steel ball, is drawn towards the magnet, thus reducing the air gap, the quantity of light detected by the phototransistor decays. The electronic control (compensation) circuit interprets this as a prompt to decrease the winding current and so let gravity restore the float to its equilibrium position.

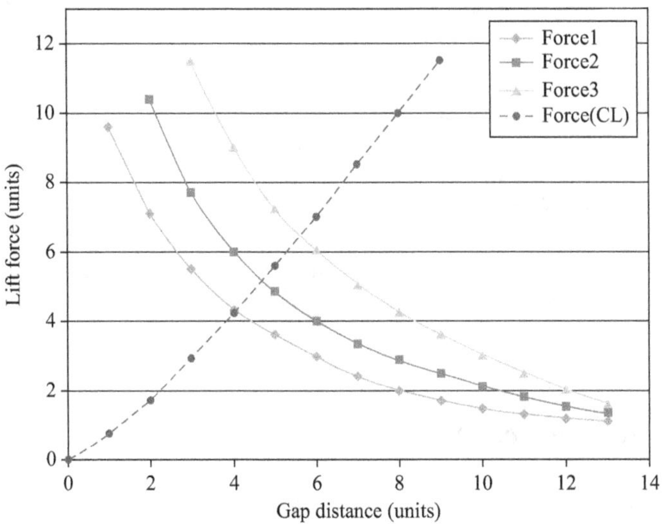

Figure 4.10 Magnetic lift force for active magnetic suspension system

The simplest form of effective compensation is one in which rates of change of position of the float are incorporated. A circuit that achieves this is shown in Figure 4.11, with signal derivatives generated by R_1, C_1 and R_2, C_2 [8]. The circuit of Figure 4.11 works by comparing the signals from the detectors at the first op-amp where a voltage proportional to the difference or 'error' is formed. The difference signal is then sent through a compensation network that essentially acts as a high-pass filter, allowing rapid changes in the input signal to pass more easily than slow changes – thus differentiating it. This is required to stabilise the control loop as suggested earlier. Without compensation, objects would just oscillate close to the electromagnet, due to lack of damping. The signal is then amplified to its original amplitude at transistor T_1, since the compensation network produces signal attenuation. The output of T_1 feeds the power transistor T_2, which controls the electromagnet current. The phototransistors used to detect the IR radiation from the light-emitting diode (LED) are not too critical to the system operation, as long as they are reasonably well matched in frequency sensitivity and the detection/emission angles are not too narrow.

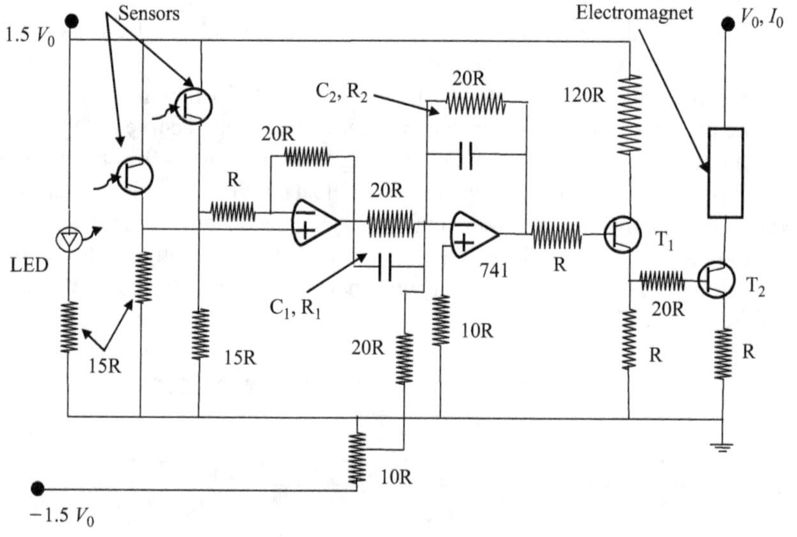

Figure 4.11 Typical analogue control circuit for a magnetic suspension system

The magnetic suspension arrangement described above displays all the essential features of an active magnetic bearing (AMB) [14–16], and we will take a look at two practical examples in section 4.3.

4.3 Oil-free magnetic bearings

While advances in magnetic materials have seen major implications for magnetic devices, components and systems, recent progress in the development of AMBs has arguably more to do with the growing computational power and speed of microprocessors. Dramatic improvements in the reliability of AMBs has also meant the emergence of applications, such as in remote pipeline pumps and in jet engine rotors, that would otherwise have been unimaginable not too long ago.

4.3.1 AMBs

In the AMB the float (see Figure 4.1) is cylindrical in shape (shaft) and encircled by a stator formed from several electromagnets – not unlike the geometry of an electric motor (Figure 4.12). The shaft is suspended within the stator by radial magnetic forces generated by the electromagnets. Two bearings at opposite ends of the spinning shaft maintain its radial position while a thrust bearing does the same axially. Patents for AMBs date back to the 1940s and an AMB first appeared in a commercial product in the 1970s. At this time they were used in very specialized roles – for example, in turbines and compressors and in high-speed centrifuges for purifying radioactive isotopes. The scope of applications began to expand as the control electronics evolved for analogue forms described in the previous section to digital alternatives. In digital technology a controller generally comprises anti-aliasing filters, A/D convertors, a digital signal processor (DSP) and pulse width modulation (PWM) devices. This technology is well rehearsed in the literature [17].

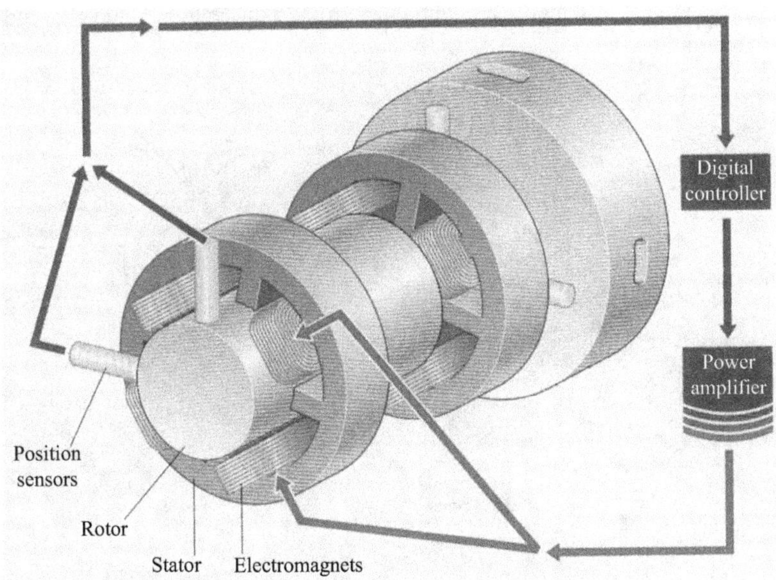

Figure 4.12 AMB showing sensors and digital control scheme

The filters minimise error signals associated with high-frequency 'noise' emanating from the detectors, and sampling errors. Once this filtering is performed, the resultant signal from orthogonally located sensors is sampled by the A/D converter for processing in the DSP [18]. There an output proportional to the shaft off-set is formed that is subsequently used to generate the necessary winding current to correct the error. The correction is supplied to the power amplifiers, which drive the electromagnets, in the form of a pulse width modulated signal, with increasing pulse width equated to rising current demand. In the scheme shown in Figure 4.12, two power amplifiers per radial bearing are required to feed orthogonally located electromagnets so that attractive restoring forces can be generated to accommodate all movements of the shaft. An advantage of this digital control process is that the 'amplifiers' are little more than high-voltage switches that respond at high frequency to the PWM drive signal. The result is high-precision bearing control than is available by analogue means. The presence of digital control embedded within the bearing in a compact integrated electronic package means that in addition to its basic suspension role, control of magnetic stiffness can be introduced to maintain shaft position for varying loads, and diagnostics can be implemented to anticipate problems before irreversible breakdown occurs, which is not easy to do with conventional bearings.

In summary, contactless operation gives AMBs several advantages over mechanical bearings. An electric motor incorporating such bearings is depicted in Figure 4.13. With frictionless bearings there is no need to provide lubrication, so the motor is in principle devoid of wear, which gives it a very long operational life. It can also function at very high speed −50,000 revolutions per minute (rpm) or more – with no bearing restrictions. Finally, lack of a lubrication regime for devices employing AMBs, gives them eco-friendly credentials, an important feature in today's world of 'green' imperatives.

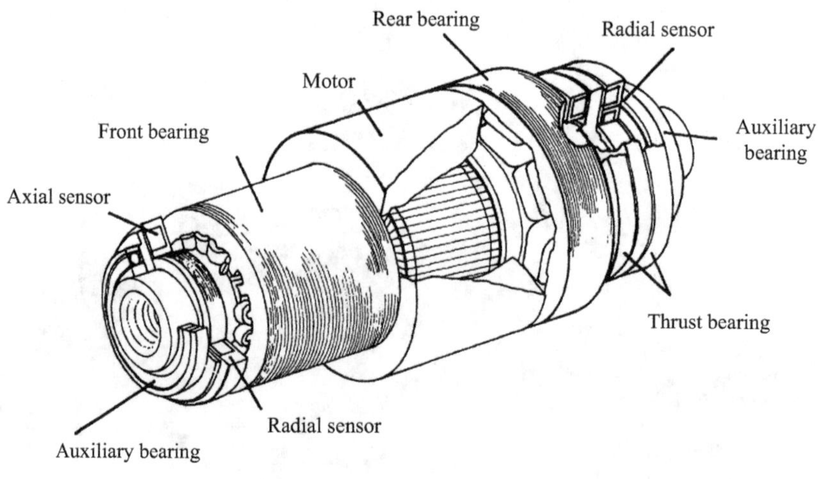

Figure 4.13 Electrical motor with oil-free frictionless bearings

4.3.2 Vertical axis wind turbines with AMB

A seemingly obvious candidate for the adoption of AMB technology is the wind turbine, which is often remotely located, and ideally should require low maintenance and be long-lasting. However, for the ubiquitous three-bladed horizontal axis wind turbines (HAWTs) this has not happened, since the load on the bearings, particularly for large 1 MW machines, is too intense and variable to be cost effectively accommodated in an AMB. Consequently, attention has focused on vertical axis wind turbines (Figure 4.14) for the introduction of frictionless bearings, since the load on the bearing, clearly visible in Figure 4.14, is relatively uniform on the thrust element and much less uneven for the radial bearing, particularly for multiple aerofoil blade types as shown. This 'aerodynamic lift type' turbine consists of two or more blades usually – but not always – mounted vertically on cross-beams attached to a vertical rotating shaft. This design of wind turbine was patented in 1931 by Georges J.M. Darrieus [1888–1979], a French aeronautical engineer. An alternative stronger and stiffer blade with a helical twist is shown in Figure 4.14. The design differs fundamentally from the more familiar cup-shaped wind collectors used on anemometers. These rely on 'drag' as do the S-shaped wind interceptors, invented by Sigurd J. Savonius in 1922. Interestingly, the Darrieus type is theoretically just as efficient as the propeller type in a constant wind speed. However, this steady breeze efficiency is rarely realised in practice due to the physical stresses and limitations imposed by wind speed variation.

In the recent past, there has been a rapid growth in the availability of, at tolerable cost, high coercive force permanent magnet materials, which have resulted in the technology of permanent magnetic suspension being expanded into wind-turbine applications. AMBs are preferred for moderate size wind turbines (10–100 kW) since,

Figure 4.14 4 KW vertical axis wind turbine with magnetic bearing

as we have seen, they provide controllable stiffness and damping. Also load capacities are compatible with such machines. They have procured the following specific advantages over conventional bearings:

- Starting wind speeds are reduced by the adoption of magnetic suspension due to reduced bearing friction, which also means turbine efficiency is significantly improved, since power output is not wasted in generating heat within the bearings.
- Magnetic suspension has radically changed the traditional wind-turbine housing design by abandoning the use of oil-film roller bearings. These mechanical bearings depend on careful lubrication and sealing to procure long service life, high-impact resistance and acceptable reliability. This requirement considerably influences design choices. The introduction of AMBs not only reduces bearing cost and maintenance, but also reduces the downtime of the wind turbine and therefore, improves the overall efficiency of the system. A key element to this benefit is the choice of generator, and in particular the adoption of a generator that is synchronous and inverted. The synchronous aspect means that the frequency of the generated electricity is governed by the rotational speed of the armature, while the inversion implies that the windings in which the electrical currents are induced are stationary, with permanent magnets embedded within the spinning armature. Consequently, no commutation, or slip-ring electrical connection, is required, which eliminates a notorious source of wear and breakdown in generating systems.

A schematic showing the key elements of the generator including AMBs and back-up mechanical bearings for a moderate capacity VAWT is presented in Figure 4.15. The device offers magnetic suspension of a wind turbine with five degrees of freedom using radial and axial magnetic bearings. The upper frictionless bearing (A) is a radial AMB, not unlike the device described in the previous section while the weight of the turbine is counteracted by the frictionless thrust bearing at (B and F). This bearing is the most important part of the VAWT magnetic suspension system. The generator stator (C) and armature (D) is shown located between the AMBs. The turbine blades (not shown) are presumed to be attached to the shaft at (H). The control system and electrical interconnections to the AMBs are also omitted for clarity – but it is not unreasonable to assume that the electronics would be located in a weather proof capsule attached firmly below the generator housing. For a remotely located wind turbine, it is very unlikely that back-up roller bearings would not be provided as shown at (E) and (G).

For VAWTs exhibiting peak output powers in the less than 10 kW range, AMBs have been developed in which the bearings are formed from permanent magnets in repulsion. The result is virtually maintenance-free devices. This advance has been made possible by the advent of high coercive force materials such as NdFeB (neodymium iron boron). Neodymium magnets are currently the best choice for these bearings because they provide high coercive force and high energy at low cost. However, in some environments NdFeB is susceptible to temperature instability, and resistance to magnetic energy erosion becomes weakened. The material is inferior

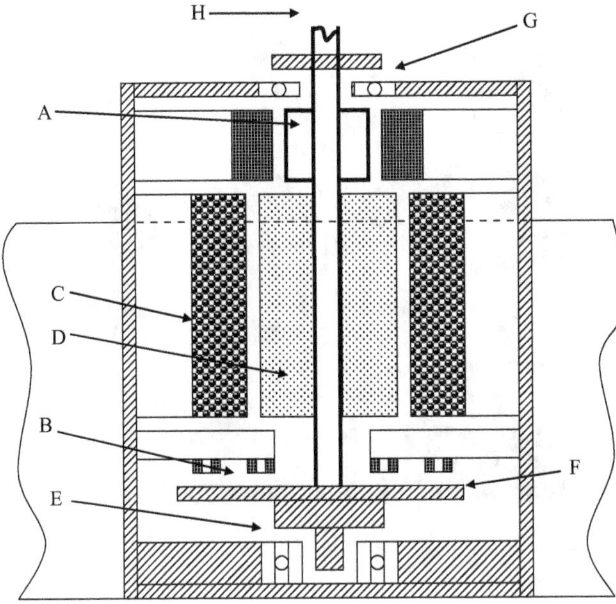

Figure 4.15 *Magnetic suspension system of vertical axis wind turbine: (A) radial AMB, (B) thrust AMB, (C) generator stator, (D) generator armature, (E) lower back-up thrust bearing, (F) thrust plate, (G) upper back-up bearing and (H) to aerofoils*

to samarium–cobalt (SmCo) in this regard. The typical NdFeB Curie temperature (at which the material becomes essentially non-magnetic) is 312°C. Below this level NdFeB exhibits a reversible temperature coefficient of remanence of −0.13/°C. This means a typical NdFeB working temperature is below 80°C. Obviously, alloys are being investigated and methods of manufacture are being evolved to procure better temperature stability and erosion resistance, to yield frictionless bearings with wider operational ranges.

Interestingly, advances [19] in the magnetic energy capability of new rare earth magnetic materials has encouraged engineers to contemplate very large (>1 MW) VAWTs employing active bearings incorporating powerful permanent magnets (see Figure 4.16). With frictionless bearings, such a machine can start turning and generating electricity in very low wind speeds (3 mph). They also work well in high wind speeds of over 80 mph whereas comparable HAWTs are designed to shut down automatically to avoid damage. It is claimed that the 'maglev' VAWT will exhibit 20% greater efficiency than conventional alternatives of similar power level.

In principle, the potential to scale up magnetic wind turbines is enormous. Researchers and engineers have postulated that 1 GW maglev VAWT farms are feasible – enough power to supply around 750,000 homes.

Figure 4.16 Vertical axis wind turbine on levitated bearing

4.4 Linear bearings in transport

Concurrently with research into maglev bearings that, as we have seen, proceeded strongly in the 1970s on the back of developments in transistors, integrated circuits and digital control electronics, maglev transport systems were attracting equal attention. As with AMBs, these technologies allowed designers to create active electromagnetic levitation (EML – including suspension) in linear geometries, by incorporating, into the moving vehicle, reliable integrated sensors feeding compact control electronics and electrical power circuits. Not surprisingly, linear levitation and suspension systems of this description are essentially 'unwrapped' AMBs but with the fixed permanent magnet rail replacing the stator, and the electromagnets on the moving vehicle replacing the armature (Figure 4.17). Apart from some geometrical niceties associated with magnetic circuit layout, the design equations are essentially those spelled out in section 4.2.

The basic configuration of an EML rail system is shown in the schematic of Figure 4.17. The traction mechanism is not shown but usually takes the form of a linear synchronous motor [8]. Suspension forces are induced between the onboard electromagnets and the permanent magnet track in much the same way as for an equivalent AMB, but as emphasised earlier in this chapter the scheme is intrinsically unstable. Consequently, for vertical stability, feedback to the electromagnets is employed based on sensing of the gap between the vehicle and the track. In early prototypes using conventional electromagnets and ferrite rails the gap was maintained at about 1 cm. Later superconducting systems, as we shall see in Chapter 6, can maintain much larger gaps of the order of 5–10 cm [6]. Lateral stability is usually achieved mechanically by simply including guide wheels on the extremity of the flange carrying the electromagnets.

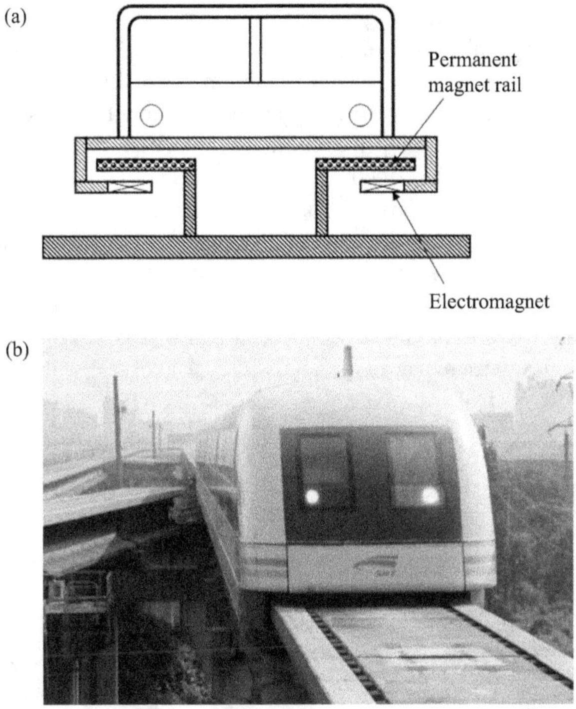

Figure 4.17 Linear maglev transport system

Finally, it is interesting to note that for passenger transport, suspension systems as shown in Figure 4.17 are preferred to levitation systems using permanent magnets [8], because leakage flux is very much lower. Consequently, the need for magnetic shielding of the passenger compartments is largely eliminated with significant benefits to costs.

4.5 Summary

Magnetic levitation technology, which entails incorporating a magnetic float within a magnetic circuit, is developed and discussed in this chapter. To appreciate the levitation issues, it is first necessary to have an understanding of linear and non-linear magnetic circuits and how they are analysed and modelled. The tools required to acquire the necessary level of comprehension are provided in section 4.1. The dynamic rules and equations governing free floats levitated or suspended by magnetic fields are addressed in section 4.2 with stability issues debated in section 4.2.2.

Practical examples that illustrate the range and effectiveness of the technology of magnetic levitation are described in section 4.3 and 4.4. In the former, the essential elements of AMBs, as used in electric motors, are introduced. Also a state-of-the-art – in terms of size – AMB as used in a modern vertical axis wind turbine is dissected in section 4.3.2, with the description giving prominence to the exotic materials required to achieve air bearings of this power level. The general public is probably most aware of magnetic levitation through the recent developments on high-speed maglev trains. The range of levitation geometries incorporated into maglev systems is extensive and well covered in the literature. In section 4.4, a geometry that is not untypical of the genre is illustrated and described.

In a coming era of power derived from renewable sources, as intimated in Chapter 1, it seems very likely that the available power will be rationed and expensive. Under such circumstances, engineers are likely to discover that the achievement of high efficiency in electrical machines will be a paramount requirement of users. In the industrial and transport arenas, it seems clear that AMBs and maglev will become critical technologies particularly as materials science evolves, as it surely will.

4.6 References

[1] *Amperes: 3D magnetic field simulation software*. Available from www.integratedsoft.com/products/amperes/default.aspx

[2] *ANSYS Maxwell*. Available from www.ansoft.com

[3] Kraus, J.D. *Electromagnetics*. London: McGraw-Hill Book Co.; 1988

[4] Grant, I.S., & Phillips, W.R. *Electromagnetism*. Chichester: John Wiley & Sons Ltd; 1990

[5] Livingston, J.D. *Rising force*. Cambridge, MA: Harvard University Press; 2011

[6] Moon, F.C. *Superconducting levitation*. New York, NY: John Wiley & Sons Inc.; 1994

[7] Meisenholder, S.G., & Wang, T.C. *Dynamic analysis of an electromagnetic suspension system for a suspended vehicle system*. U.S. Federal Railway Administration Report, No. FRA-RT-73-1, 1972

[8] Jayawant, B.V. *Electromagnetic levitation and suspension techniques*. London: Edward Arnold Publishers Ltd.; 1981

[9] Maslen, E.H., & Schweitzer, G. (Eds.). *Magnetic bearings*. Berlin: Springer-Verlag Ltd.; 1988

[10] Rhodes, R.G., & Mulhall, B.E. *Magnetic levitation for rail transport*. Oxford: Clarendon Press; 1981

[11] Thaler, G.J., & Brown, R.G. *Analysis and design of feedback control systems*. New York: McGraw-Hill Book Company Inc.; 1960

[12] D'Azzo, J.J., & Houpis, C.H. *Linear control system analysis and design*. London: McGraw-Hill International Book Company; 1981

[13] Brogan, W.L. *Modern control theory*. London: Prentice-Hall International (UK) Ltd.; 1991

[14] Jayawant, B.V., Sinha, P.K., Wheeler, A.R., Whorlow, R.J., & Willsher, J. 'The development of a one ton magnetically suspended vehicle using controlled DC electromagnets'. *IEE Proceedings*. 1976;**123**(9):941–948

[15] Jayawant, B.V., Sinha, P.K., & Aylwin, D.G. 'Feedback control systems for DC electromagnets in passenger carrying vehicles'. *International Journal of Control*. 1976;**24**(5):627–639

[16] Sinha, P.K. *Electromagnetic suspension: dynamics and control*. London: Peter Peregrinus Ltd.; 1987

[17] Comer, D.J. *Electronic design with integrated circuits*. Reading, MA: Addison-Wesley Publishing Company Inc.; 1981

[18] Richards, G. 'Magnetism on a roll'. *IET Engineering & Technology*. March 2010;**5**(4):37–39

[19] Liu, S. 'Magnetic suspension and self-pitch for vertical-axis wind turbines, fundamental and advanced topics in wind power'. In R. Carriveau (Ed.), *InTech*, 2011. ISBN: 978-953-307-508-2. Available from http://www.intechopen.com/articles/show/title/magnetic-suspension-and-self-pitch-for-vertical-axis-wind-turbines

Chapter 5
Electrostatic suspension

> We live in a society exquisitely dependent on science and technology, in which hardly anyone knows anything about science and technology.
>
> *Carl Sagan*

5.1 Suspension of metallic objects

The evolution of frictionless bearings in instruments and machines, including the contactless manipulation of objects, has been advanced predominantly by adopting electromagnetic methods, rather than electrostatic methods, to create levitation and suspension systems as described in Chapter 4. This is partly because, even more so than conventional magnetic levitation, which is only conditionally stable as we have already discovered, electrostatic suspension has generally been considered to be a particularly fraught area of technology because the Earnshaw's rules influence stability more acutely. Nevertheless, electrostatic suspension offers some significant advantages over magnetic techniques.

First, it is possible to suspend a wide range of different materials from metals to semiconductors to dielectrics. Magnetostatic systems, or more correctly permanent magnet systems, are essentially limited to ferromagnetic materials. In semiconductor manufacturing and processing, the contactless and friction-free transportation and manipulation of wafers in ultra-high vacuum and ultra-clean chambers has become an essential tool of reliable production. This capability is achieved through electrostatic suspension methods. Second, electrostatic suspension generates very little heat in comparison with magnetic techniques, making it much more suitable to vacuum applications. Finally, electrostatic electrodes can readily be made to conform to the requirements of micro-scale planar structures, because at this scale electrostatic energy densities are much higher than in magnetic alternatives and hence suspension force levels become increasingly effective. Furthermore, the burgeoning requirements of the evolving planar technology sector, as represented by micro-electro-mechanical systems (MEMS), have given a considerable boost to the search for frictionless actuation.

On the debit side, as has already been suggested above, stable suspension is particularly difficult to achieve with electrostatic fields because of relatively high voltages and poor levels of stiffness. An active and sophisticated intervention strategy to maintain control is mandatory. In addition, on-going developments in electrostatic suspension have been hindered by the fact that the forces available to make objects float are often insufficiently strong to do so effectively at normal atmospheric pressure because air breakdown limits the usable voltage levels, as we shall see. Vacuum applications are not so restricted and in this technology area progress has been steady. These issues will be explored in the ensuing sections.

The essential features of electrostatic suspension are encapsulated in the simple schematic drawing in Figure 3.13 introduced in section 3.3. The basic force relationship associated with this electrode configuration is provided by (3.25). However, in practice, the two circular stator pads in Figure 3.13 usually become elements in a circumferential array of metal pads or fingers within a stator disc, below which floats a metal or non-magnetic dielectric disc. For any given pair of stator electrodes, in such an assembly, (3.25) is still applicable. The situation is illustrated in Figure 5.1a, which

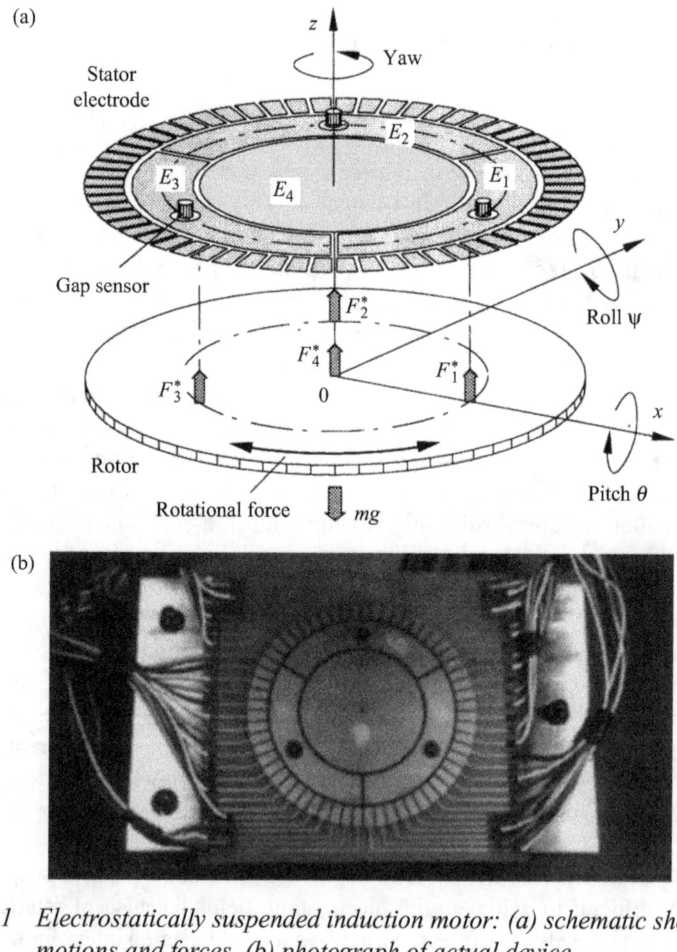

Figure 5.1 Electrostatically suspended induction motor: (a) schematic showing motions and forces, (b) photograph of actual device

provides a drawing of an electrostatically suspended induction motor. A photograph of the actual device is inserted below it in Figure 5.1b [1,2].

As we have already observed in Chapter 3 the stator electrodes together with the rotor (flotor in some papers) can be modelled using variable capacitors, and when these electrodes are coupled to a multiplicity of neighbours as in the case of an array of fingers/pads (see in Figure 5.1), they form a ladder-like electrical circuit. If the flotor is composed of a highly conducting material, then it is reasonable to assume that its surface is an equipotential surface. Furthermore, if the flotor was originally uncharged then it remains so when in suspension close to the stator.

5.1.1 Air breakdown constraint

Experiments performed on aluminium discs of varying thickness indicate that electrostatic suspension of a metal object is severely limited by electric field breakdown of the air in the gap between the stator and the float [3]. Results presented in Table 5.1 show the stabilised voltages and gaps for 5, 10, 15 and 20 mm thick discs suspended in vacuum by a disc-shaped stator divided into four sections. The final row in the table displays the electric field in the gap at equilibrium by making the approximation (3.18) that the electric field in the gap is just the voltage divided by the gap width. It is clear that, given air breakdown at normal pressure and temperature occurs at a field stress of 24 kV/cm, none of the discs employed in the experiment could be suspended in air. The disc thickness has to be reduced to 1 mm to procure suspension without breakdown.

Table 5.1 Gap fields for aluminium discs at equilibrium

Thickness of disc (mm)	5	10	15	20
Mass of disc (g)	106	211	334	424
Gravity/unit area (kN/m^2)	1.32	2.63	4.17	5.29
Electrode voltage at equilibrium (kV)	1.34	1.74	2.28	2.68
Gap at equilibrium (mm)	200	200	210	250
Electric field in gap (kV/cm)	67	87	109	107

5.1.2 Optimum suspension condition

Although the flotor in an electrostatic suspension system is generally kept charge-free by careful fabrication and handling, it is not necessarily at zero potential. Unfortunately, this potential can influence the efficiency and stability of the flotation system by creating uncontrolled and unpredictable forces to ground. It is important, therefore, to seek to minimise float potential. How this can be done is best demonstrated by considering the equivalent electrical circuit (see Figure 5.2) for a multi-electrode suspension arrangement.

In Figure 5.2 the stator electrodes (n in number) are fed by voltages $V_1, \ldots, V_n$. The capacitances between each electrode and the metal float are represented by the capacitance symbols that have values $C_1, \ldots, C_n$. The float is presumed to exhibit

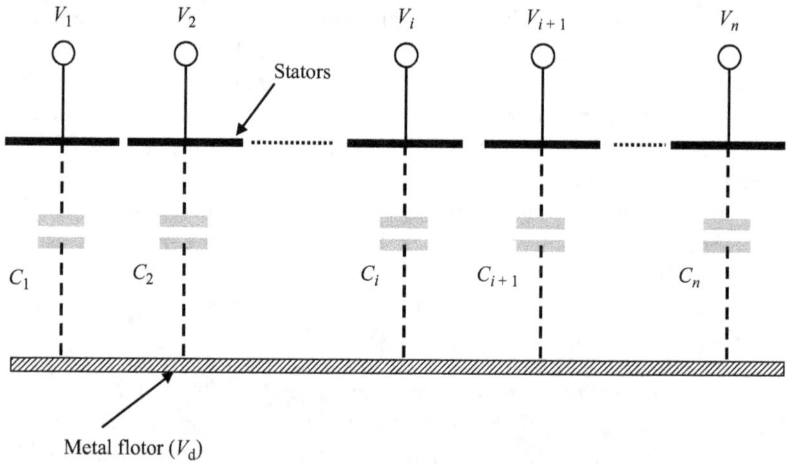

Figure 5.2 Circuit model of electrostatic suspension system

a potential of magnitude V_d. If we presume, reasonably, that the electrodes are smooth, flat and highly conducting, then the boundary condition at each electrode (see Chapter 2) requires that:

$$D_n = \sigma \tag{5.1}$$

where D_n is the electric flux density normal to the electrode surface and σ is the surface charge density in C/m^2. Hence the normal electric field E_n is given by

$$E_n = \frac{\sigma}{\varepsilon_r \varepsilon_0} \tag{5.2}$$

But E_n represents the voltage gradient between a given electrode and the float. That is

$$E_n = -\frac{\partial V}{\partial n} \tag{5.3}$$

where ∂n denotes an elemental distance in a direction normal to the electrode surface. Hence we can write

$$\frac{\sigma}{\varepsilon_r \varepsilon_0} = -\frac{\partial V}{\partial n} \tag{5.4}$$

Consequently, referring to Figure 5.2, for the ith electrode of area A_i (say), the charge Q_i on the electrode and on the float immediately below is $Q_i = \sigma_i A_i$ and for a simple capacitor (C_i) we also have:

$$\frac{\partial V}{\partial n} \approx \frac{V_i - V_d}{d_i} \tag{5.5}$$

Hence from (5.4):

$$\frac{Q_i}{V_i - V_d} = \frac{\varepsilon_r \varepsilon_0 A_i}{d_i} = C_i \tag{5.6}$$

(see Chapter 2). If the C_i's are known for a given device and if the applied voltages V_i are preset, then the computation of levitation forces, which depend on the Q_i values, implies that the value of V_d for the applied electrode voltages has to be predeterminable. In other words, V_d should be expressible as a function of V_i and C_i for $i = 1, 2, \ldots, n$.

An expression for V_d for the general multi-electrode geometry can be constructed by examining current flows in the ladder circuit, and this is made easier by introducing gap leakage resistances R_i in parallel with the gap capacitances, as shown in Figure 5.3.

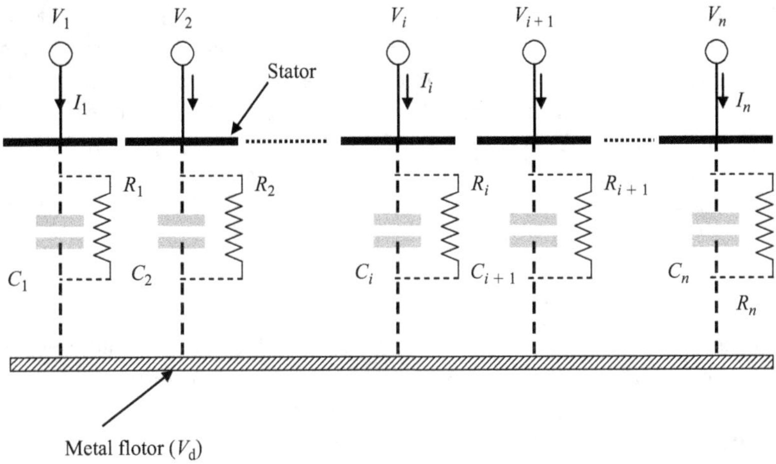

Figure 5.3 Circuit model of electrostatic suspension system including leakage

The currents in the capacitive branches of Figure 5.3, namely I_i for all $i = 1, 2, \ldots, n$, are

$$I_i = j\omega\left[\frac{C_iR_i + 1}{R_i}\right](V_i - V_d) \tag{5.7}$$

But Kirchoff's laws require that:

$$I_n = -\sum_{i=1}^{n-1} I_i \tag{5.8}$$

so that when these two equations are combined we obtain:

$$V_d = \frac{\sum_{i=1}^{n} \frac{V_i}{Z_i}}{\sum_{i=1}^{n} \frac{1}{Z_i}} \tag{5.9}$$

where the impedance at the ith electrode is

$$Z_i = \frac{R_i}{j\omega C_iR_i + 1} \quad \text{for } i = 1, 2, \ldots, n \tag{5.10}$$

Note that when R_i can be considered to be infinitely large, (5.9) becomes:

$$V_d = \frac{\sum_{i=1}^{n} V_i C_i}{\sum_{i=1}^{n} C_i} \tag{5.11}$$

An appreciation of the relevance of (5.9) to the electrostatic suspension problem represented by planar multi-electrode devices, such as the one shown in Figure 5.1a, can be acquired by considering two symmetrical arrangements in which $Z_i = Z$. First, if the electrodes are fed in phase with equal magnitude voltages ($V_i = V$), then it is not difficult to deduce from (5.9) that $V_d = V$. Consequently, there is no field between the electrodes and the float, and no suspension force is possible. On the other hand, if the electrode voltages are of equal magnitude ($V_i = V$) and in anti-phase ($V_i = -V_{i+1}$), (5.9) yields $V_d = 0$. This choice results in optimum field strength between the electrodes and the float, and maximum suspension force strength. Each electrode pair in this case operates in the manner suggested by Figure 3.13. In general, the necessary condition for $V_d = 0$ and optimum force is

$$\sum_{i=1}^{n} \frac{V_i}{Z_i} = 0 \tag{5.12}$$

5.1.3 Force equations

The forces associated with electrostatic suspension of a float in a multi-electrode planar structure can generally be expressed in simple closed-form equations particularly if the arrangement displays a high degree of symmetry. This will be demonstrated in this section by examining the four-electrode system depicted in Figure 5.4, which is not untypical of the genre, except for the small number of electrodes. In these days of readily available computer simulation packages for solving electromagnetic boundary value problems [4,5] it might seem pointless to expend

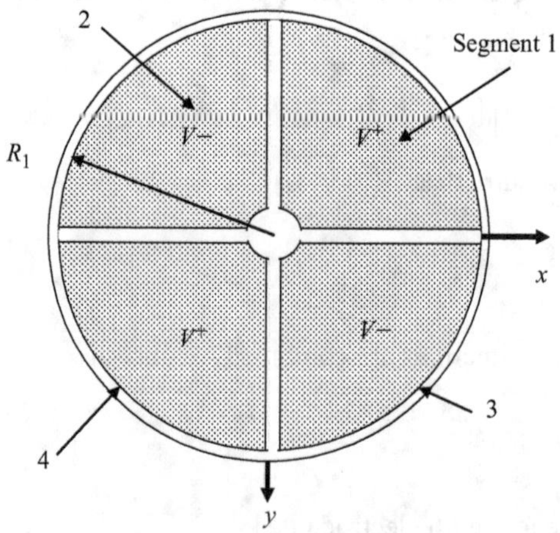

Figure 5.4 Stator electrode pattern employed in comparison of measurement and theory (R_1 = 47.5 mm; inner aperture radius R_2 = 12.5 mm)

effort on the mathematical construction of closed-form equations. But this would not be true. MEMS devices are often embedded within complex multi-component systems that themselves are largely designed by computer modelling. These simulations tend to involve the iterative addressing of all elements of the system until an equilibrium solution is achieved. Systems modelling of this class would become extremely inefficient and computationally time consuming if each time a MEMS component had to be adjusted the simulation was forced to access a 'full-blooded' finite element solver. Needless to say, therefore, the effectiveness of such system design packages is critically dependent upon the capability of representing MEMS devices by easily evaluated closed-form equations.

In Figure 5.4, the four electrodes of the electrostatic suspension system are quadrant shaped, each with area 14.5 mm^2. They form part of a planar circular geometry supporting a circular disc of 95 mm diameter. The inner aperture of the electrode structure is 25 mm in diameter. The electrode segments are fed with equal and opposite voltages in a symmetrical manner, which means, according to (5.11), that the float is at zero potential in its equilibrium position [6].

Given that the forces on the disc can be equated to the rate of change of energy (see Chapter 3), then the following restoring force equation, in vectorial form, for the ith electrode can be constructed for the case of a lateral translation (x_s) of the disc away from equilibrium:

$$\mathbf{F}_i(V_i, x_s, g_i) = -\frac{\partial W_e(V_i, x_s, g_i)}{\partial x_i}\mathbf{a}_x - \frac{\partial W_e(V_i, x_s, g_i)}{\partial z_i}\mathbf{a}_z \tag{5.13}$$

where $\mathbf{a}_x$, $\mathbf{a}_y$ and $\mathbf{a}_z$ are unit vectors along x, y and z, g_i is the gap between the float and the ith electrode, and W_e is the electrical energy stored in the ith gap capacitor (see Chapter 3) given by

$$W_e(V_i, x_s, g_i) = \frac{1}{2}C_i(V_i - V_d)^2 \quad \text{(Joules)} \tag{5.14}$$

The second term in (5.13) is the vertical restoration, or suspension, force. When combined with (5.14) this force can be expressed as

$$F_{z,i} = (V_i - V_d)\left(\frac{1}{2}(V_i - V_d)\frac{\partial C_i}{\partial z} + C_i\frac{\partial V_d}{\partial z}\right) \quad \text{(Newtons)} \tag{5.15}$$

This equation is quite general in its application but has to be applied over all electrodes of the levitator under investigation to determine the full suspension force on the float (F_z). For the structure in Figure 5.4, for example this involves a summation over all four electrodes [6]. The procedure and the resultant equation are presented in Appendix II.

Lateral displacement of a float relative to a stationary electrode will clearly result in the development of an electrostatic force 'pulling' the disc back to the equilibrium position. This force is expressed by the first term on the right-hand side of (5.13) and is directed along $-x$ for a displacement x_s. Hence using (5.14) the lateral restoration force is

$$F_{x,i} = \frac{\varepsilon_r \varepsilon_0}{g_i}(V_i - V_d)\left(\frac{1}{2}(V_i - V_d)\frac{\partial A_i}{\partial x} + A_i \frac{\partial V_d}{\partial x}\right) \tag{5.16}$$

A similar equation is generated for displacement y_s in y-direction. The total restoring force for the four-electrode system shown in Figure 5.4 has been fully developed in reference 6. The details of the derivation need not concern us here. However, helpfully the paper provides a graphical representation of the solution and has compared it with measurement (Figure 5.5). Except for very small and very large disc movements the agreement between the theory and the measured result is well established. It is clearly more than good enough to permit the confident embedding of an algorithm based on this theory within a MEMS modelling and design package, possibly incorporating one or more four-electrode suspension devices.

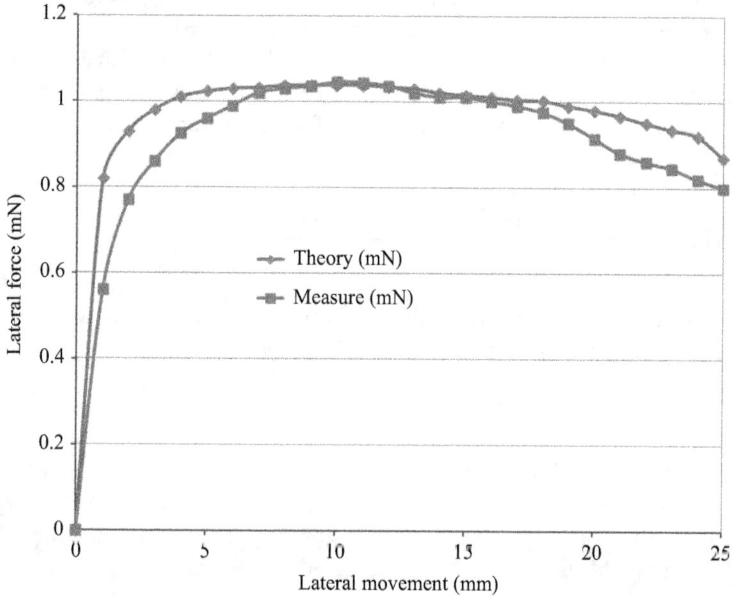

Figure 5.5 Lateral force as a function of lateral displacement for a four-electrode suspension system

5.1.4 Suspension and propulsion

Many components or products in the electronics sector of industry require a high degree of cleanliness during manufacture and hence great care has to be taken to avoid surface contamination by dust particles, which can greatly diminish yield. Significant improvements in this regard are afforded by contactless processing, but even so, difficulties continue to be encountered. These are due, first, to the accumulation of charge, and hence contaminants, on the surface of the transported object. This can be minimised by ensuring that the object remains at zero potential (see section 5.2.2). Second, if sensitive integrated circuit elements are being transported contactlessly these can be susceptible to damage by the levitation field. Fortunately, the difficulty can generally be avoided [7] by ensuring that such devices are shielded from the suspension fields by the float itself. Finally, contaminants in the air, which are 'picked up' by the electrodes, can become charged and then transferred directly by repulsion from the electrodes to the float. The solution lies in coating the electrodes with a thin insulating film [7].

A schematic of a system designed to linearly propel [7], in a contactless manner during processing, a 3.5 in. diameter aluminium disc, as used in computer hard drives, is shown in Figure 5.6. The drive system is 95 mm wide – just larger than the diameter of the disc. The rectangular electrode propulsion plate is formed from two printed copper ladders (denoted A and B in the figure) separated by a 2 mm gap. The electrode strips are 3.5 mm wide. Suspension of the 14 g disc is achieved by applying 1,000 V (DC) to the electrodes above, and overlapped by, the disc. Furthermore, in order to ensure that the disc potential remains zero, in accordance with (5.11), the voltages are applied in four quadrants (not unlike Figure 5.4) with voltage inversion implemented between adjacent regions. The four displacement sensors provide information to a

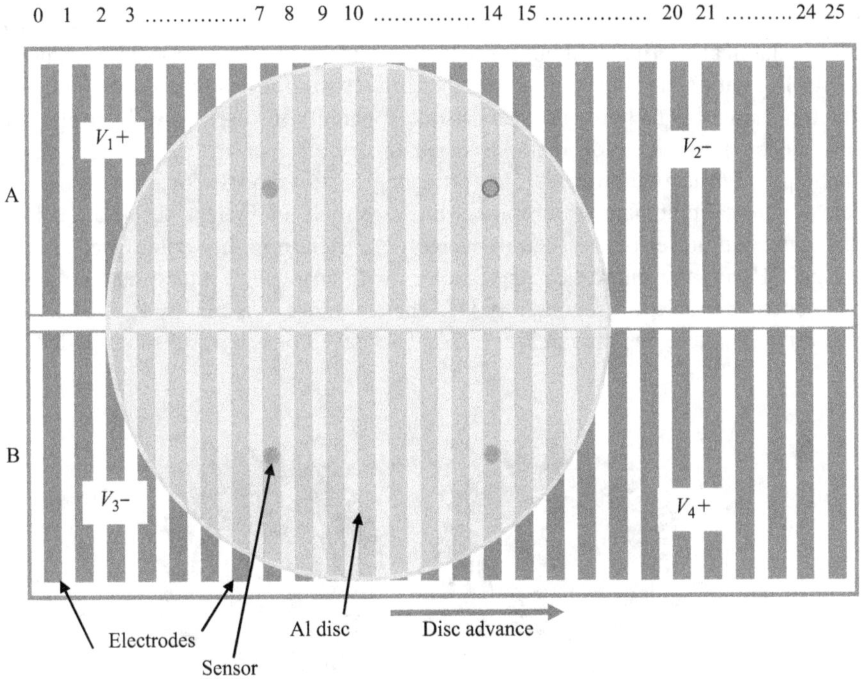

Figure 5.6 Relative position of disc with respect to suspension and propulsion electrodes, in contactless transportation system (disc is shown semi-transparent to clarify concept)

controller that directs error voltages to the active electrodes in order to ensure that the lateral disc movements are within prescribed limits, that the floatation gap is held steady, and pitch and roll movements are restricted.

Advancing the disc in the transportation direction entails sequential switching of the electrode voltages to create a component of force between the electrodes and the float that will ease the disc forward. For example with reference to Figure 5.6, at the instant shown electrodes A2–A10 are at +1,000 V, electrodes A11–A17 at −1,000 V, electrodes B2–B10 at −1,000 V and electrodes B11–B17 at +1,000 V. The applied voltages are, of course, modified by error voltages from the disc stability controller. To maintain movement to the right the driving unit of the electronic control system switches off electrodes A2 and B2 while simultaneously switching on A18 and B18. The process is repeated regularly at a switching rate that maintains forward movement at a desired speed. Note that above the disc, the electrode quadrants of positive and negative voltage are preserved during the driving process. This disc transportation concept has been validated by experiment [7,8].

5.2 Electrostatic motors with suspended rotors

While electrostatic motors of various descriptions were being discussed as far back as 1969 [9], these were not at the micro-scale of dimensions and friction associated with mechanical bearings did not present a serious issue. However, with the advent of MEMS manufacturing techniques in the early 1980s, the desire to develop micro-actuators using this technology soon became apparent [10–13]. As we have already seen, electrostatic actuation methods result in devices that are very conducive to lithographic fabrication processes. Unfortunately, it was soon discovered that micro-scale electrostatic actuators and motors are extremely inefficient because most of their generated power is absorbed in overcoming friction in the 'crude' mechanical bearings. Hence the interest in micromotors with suspended rotors, as a solution to the bearing problem, grew rapidly [1]. The technique was mainly applied to the axial-drive structure depicted in Figure 5.7. This development was also propelled by

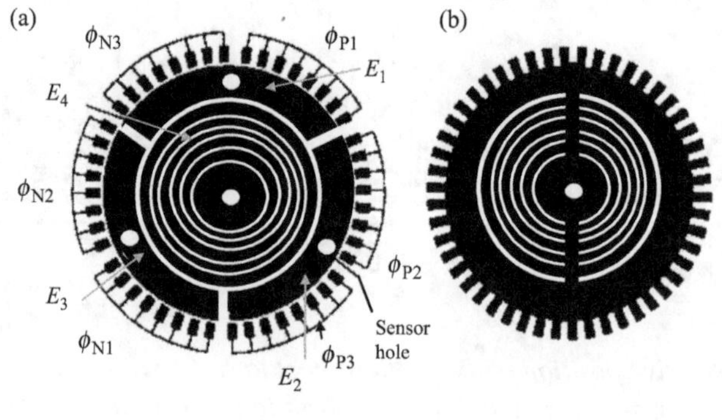

Figure 5.7 Axial-drive electrostatic micromotor: (a) stator, (b) rotor

the need to find actuators capable of being employed in a clean vacuum environment without introducing contaminants, which devices with bearings that are susceptible to wear fail to do.

5.2.1 Axial-drive micromotor

The forces responsible for the rotation of the electrostatically driven rotor in an axial-drive device, as shown in Figure 5.7, are due to the interaction between the charges on the surface of the rotor and stator (see Chapter 3). In this kind of motor the drive is usually provided by an outer ring of electrode pads around the rim of the stator interacting electrostatically with the ‘fingers’ on the rotor. Suspension and stability relies on controlled activation of four inner electrodes denoted as E_1, E_2, E_3 and E_4 in Figure 5.7a. These drive and suspension forces, as we have seen in Chapter 3, can be derived from the electrostatic field strength within the stator/rotor gap, each of which acts as a capacitive element.

The electrostatic forces providing motor action on a micro-rotor can be categorised in terms of the three basic orientations, namely axial, radial and rotational. To maximise the torque produced by the micromotor high rotational forces are desired, whereas axial and radial force components should be designed to be minimal since they are directly responsible for rotor displacement and instability. The electrostatic forces can be calculated analytically by differentiating the stored energy (U_e) in the electric field, bridging the gap between a stator and rotor finger. The differentiations are performed with respect to the acting directions of the desired forces (see (3.22)). For motors, such as that in Figure 5.7, with relatively simple rotor/stator gap geometries U_e can be determined analytically enabling closed-form force equations to be developed, as shown in sections 5.2.2 to 5.2.6.

5.2.2 Force calculations

An axial-drive micromotor, with a single-stator and a single-rotor, can be considered to comprise a multiplicity of simple parallel plate capacitors with infinitely thin electrodes, and with one of the two sets of plates rotating about a central axis. This causes the rotor/stator overlap to vary resulting in changing capacitance. The overlap is easily calculated if the fringing fields between the plates, which are generally very small, are neglected. The electrostatic force, $\mathbf{F_e}$, appearing on one of the poles can be expressed quite generally as

$$\mathbf{F_e} = -\nabla(U_e) \tag{5.17a}$$

In this context a ‘pole’ is a source or a sink of electric field and can be considered to be either a stator or rotor finger. Equation (5.17a) can be expanded to give:

$$\mathbf{F_e} = -\left\{\mathbf{a_r}\frac{\partial}{\partial r}(U_e) + \mathbf{a}_\varphi\frac{1}{r}\frac{\partial}{\partial \varphi}(U_e) + \mathbf{a_z}\frac{\partial U_e}{\partial z}\right\} \tag{5.17b}$$

on expressing the del-operator in the cylindrical coordinate system, which most easily accommodates motors of this class. Note that the axial z-direction is normal to the electrode surfaces in Figure 5.7. Also, on expanding U_e we obtain for a motor drive voltage V:

$$U_e = \left(\frac{1}{2} \frac{\varepsilon_0 V^2 A(\varphi)}{d(z)} \right) \tag{5.18}$$

where ε_0 is the permittivity of free space, V is the potential difference between the electrode finger and the rotor and $d(z)$ is the gap width which is a function of z. The overlapping area $A(\phi)$, for one rotor pole (see Figure 5.8) is given by

$$A = \frac{\varphi}{2}(R_{RO}^2 - R_{SI}^2) \tag{5.19}$$

where R_{RO} is the outer radius of the rotor, R_{SI} is the inner radius of the stator pole and ϕ is the angle defining the overlapping area of the capacitor in radians.

Note that the above formulae assume that the rotor and stator are flat plates with negligible thickness. However, while the thicknesses of the layers produced by MEMS fabrication processes in the formation of the rotor and the stator are small, they are hardly negligible. By choosing to employ the convenient assumption that the electrodes are of zero thickness means that inaccuracies in the calculated values become inevitable, since edge fields, where they exist, naturally cause the actual capacitance between rotor and stator to be marginally larger than is computed.

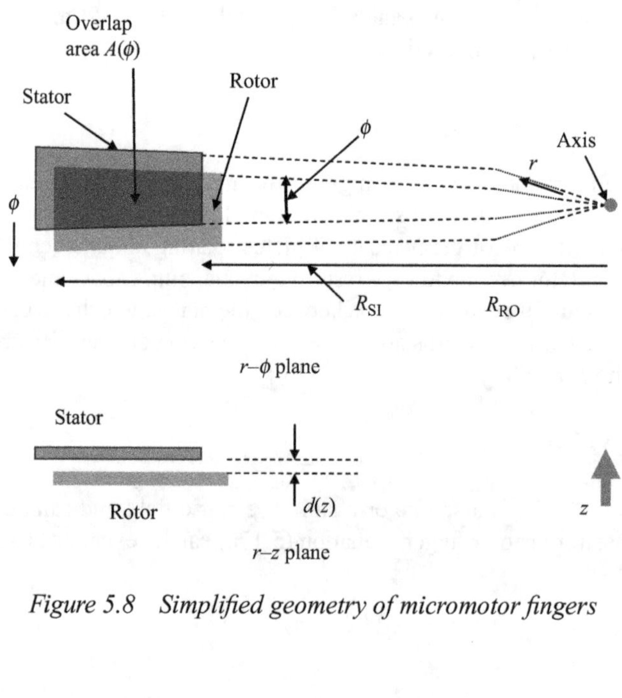

Figure 5.8 Simplified geometry of micromotor fingers

5.2.3 *Axial force expression*

Using the assumption of the parallel plate capacitor geometry the axial force per finger, F_{eA}, can be written as

$$F_{\mathrm{eA}} = -\frac{\partial U}{\partial z} = \frac{1}{2}\varepsilon_0 V^2 \frac{\varphi}{2}(R_{\mathrm{RO}}^2 - R_{\mathrm{SI}}^2)\left(\frac{1}{d^2}\right) \tag{5.20}$$

In a multi-finger axial-drive motor, several fingers may overlap simultaneously and the total axial force will be the sum of these contributions. This force contributes to tilting and rocking of the rotor and to general instability. It can be minimised for the rotor as a whole by arranging the finger design so that the force is distributed symmetrically over several electrodes.

5.2.4 *Radial force expression*

The rotor is also affected by radial forces when a mechanical disturbance causes the rotor to be displaced laterally. The radial forces for one rotor pole can be calculated by taking the partial differentiation in (5.17) with respect to the radius r and assuming R_{RO} is a function of r for small displacements. This leads to the following relation for radial force per electrode finger:

$$F_{\mathrm{eR}} = \frac{1}{2}\frac{\varepsilon_0 V^2}{d}\varphi R_{\mathrm{RO}} \tag{5.21}$$

Since, for the case of the complete rotor, there are two or more active rotor poles the total force on the rotor as a whole will be the vector sum of the individual, largely opposing, contributions. Again rotor design and excitation sequencing is usually arranged to maximise radial force cancellation, so that the resultant force on the rotor as a whole is minimised.

5.2.5 *Rotational force expression*

The rotational force, or motor force (F_{eM}) on the rotor, produces the desired motive torque of the micromotor. An analytic formula for the force exerted on one rotor pole can be obtained by taking the partial differential with respect to the angle of overlap, ϕ, of the rotor and stator (second term on the right of (5.17)), giving:

$$F_{\mathrm{eM}} = \frac{1}{r}\frac{\partial U}{\partial \varphi} = -\frac{1}{2}\frac{\varepsilon_0 V^2}{d}\frac{(R_{\mathrm{RO}}^2 - R_{\mathrm{SI}}^2)}{(R_{\mathrm{RO}} + R_{\mathrm{SI}})} \tag{5.22}$$

Because the differentiation produces a formula (5.22) with no variable that is a function of the angle of overlap, ϕ, this implies that the tangentially directed ($\mathbf{a}_\phi$) motor force is constant as long as there is a change in overlap while the rotor is turning.

The formula cannot be applied when there is either no overlap between rotor and stator, or when the overlapping area between rotor and stator does not change, since the original equation assumes parallel plates with no fringing. Therefore, there will be an initial tangential force of zero for no overlap, a constant tangential force once the rotor and stator begin to overlap and when the overlap of rotor and stator disappears, there will again be no tangential force. This force can be considered to be exerted midway along the leading edge of the rotor tooth. The behaviour is demonstrated in Figure 5.9 for a typical electrostatic micromotor with 8 rotor fingers, 12 stator electrodes and $R_{RO} = 160\ \mu m$.

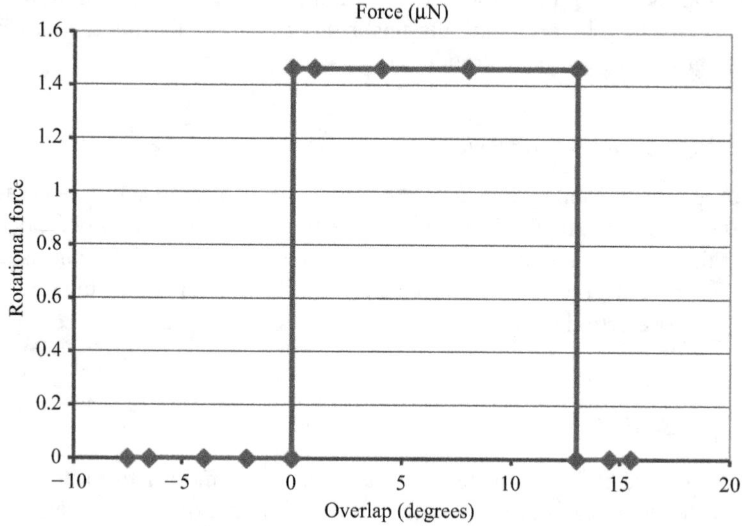

Figure 5.9 Rotational force as a function of angle of rotation (degrees) for an electrostatic micromotor

5.2.6 Rotational torque expression

The static torque produced by the electrostatic charge-induced fields between stator and rotor can be calculated by extending the equation for rotational force (5.22) to include the effective radius at which the calculated force is exerted on the rotor pole. This produces the following equation for the torque per pole:

$$T_e = \frac{1}{4}\varepsilon_0 \frac{V^2}{d}(R_{RO}^2 - R_{RI}^2) \tag{5.23}$$

Note that the factor in the force expression that places it at the mid-point of the edge of the plate has disappeared in the torque equation. This is because torque is simply the force multiplied by the radius at which that force is exerted. The total torque on the rotor is, of course, the summation of the torques developed at all active poles in a multi-finger axial-drive motor.

With this basic understanding of the primary force relationships established, we can now consider some examples of practical micromotors incorporating rotor suspension.

5.2.7 Practical axial-drive rotor

Lithographic electrode patterns for an operationally successful suspended axial-drive motor [1] are presented in Figure 5.7. The rotor and stator are etched from a 35 μm thick copper layer on a plane glass substrate. The rotor is 100 mm in diameter at the 'finger' or 'tooth' tips and it weighs 10.8 g. There are 50 teeth with width/space ratios of unity so that each tooth subtends an angle of 3.6°. The tooth depth is 5 mm. The electrodes on the stator, delivering motor propulsion, are arranged in six groups of eight with each electrode pad equal in area to a rotor tooth and located at a radial position that ensures perfect overlap. Each group represents a single excitation phase, denoted in the figure by ϕ_{Pi} and ϕ_{Ni} $(i = 1, 2, 3)$ with the voltage (~700 V) applied to the P-phases being positive and to the N-phases being negative. The sequencing ensures zero rotor potential. In each stator group the electrode pitch matches the tooth pitch on the rotor, but at the group boundaries the pitch is third larger than the standard pitch. This adjustment makes it possible to maintain a continuous driving force with simple sequencing that proceeds as follows. The electrode groups ϕ_{Pi} and ϕ_{Ni} are excited progressively in the order $i = 1, 2, 3$, for counterclockwise rotation. If the first pair to be energised is phase 1 then the rotor will move either anticlockwise or clockwise to align with the phase 1 stators. Thereafter, it moves counterclockwise if excitation shifts to phase 2 then to phase 3 and so on, in the order 1231 .... This is because at each voltage switch, the phase being activated will instantaneously exhibit a 1/3 pitch displacement between its stator electrodes and the nearest rotor teeth, thus pulling the rotor counterclockwise to re-establish alignment. The sequence 1321 ... produces clockwise rotation.

Suspension and stability is provided by a pattern of electrodes in the central area of the stator together with matching conduction areas on the rotor (Figure 5.7). The stator area is divided into four sections. These are denoted E_1, E_2, E_3 and E_4 in the figure. The outermost electrodes E_1, E_2 and E_3 are formed from the division into three 120° sectors, a metalized stator ring, which is located just inside the array of propulsion electrodes. At equilibrium these electrodes are set at 530 V but they can vary by ±40 V in the suspension control process that depends on three contactless gap sensors positioned at the centre of each electrode (see Figure 5.7). The inner electrode ring (E_4) is set to −530 V at equilibrium and also varies as much as ±40 V in the rotor controlling process. This arrangement ensures that the rotor remains at essentially zero potential. The five ring-shaped slits etched into electrode E_4 has nothing to do with suspension control but is introduced to provide a force mechanism capable of restricting lateral movement of the rotor [1].

Measurements on the suspended motor demonstrate rotational speeds of up to 60 rpm with the rotor stably suspended at a gap of 300 μm. At higher speeds the rotor becomes unstable laterally because the restriction mechanism is no longer strong enough to keep the rotor in place. The rotor accelerates to 60 rpm in 10 s, which implies that a rotational torque of the order of 8.5 μN m is being generated.

5.2.8 Suspension of dielectrics

Technological activities are evolving, associated with micro-electronics and MEMS, which increasingly demand contactless handling not just of metallised objects alluded to above but also of insulators such as glass. In liquid crystal displays, for example where the norm is the handling of glass panels by robotic arms and mechanical grabbers, handling difficulties are being experienced as a result of the predominant development trend that is for the thickness/surface area ratio of the panels to be progressively reduced [2]. Physical contact between glass and handling tools has a further disadvantage that the surface of the non-conducting glass can easily be contaminated with particles through frictional processes and electrostatic charge build-up. Consequently, contactless processing of dielectrics is becoming increasingly attractive to manufacturers.

Suspension of non-conduction objects is possible only by electrostatic means and depends on a fundamental property of dielectrics – namely polarisation. In Chapter 2, it has already been observed that the basic electric field quantities in electromagnetics, namely the electric flux density **D** and the electric field intensity **E**, are related through the equation:

$$\mathbf{D} = \varepsilon_r \varepsilon_0 \mathbf{E} \tag{5.24}$$

where ε_0 is the absolute permittivity of free space or vacuum. In general, the relative permittivity ε_r is a complex quantity accommodating polarisation, conducting, relaxation and resonance effects [14]. However, for a perfect (non-conducting) dielectric it is real at low frequencies and at DC, and we can write the above equation in the form:

$$\mathbf{D} = \varepsilon_0 \mathbf{E} + \mathbf{P} \tag{5.25}$$

Here, the vector **P** is the electric polarisation vector per unit volume of the dielectric, a classical quantity in electromagnetic theory. Without any real loss of generality we can assume that electrostatic suspension will generally be applied to perfect, homogeneous and isotropic dielectrics, in which case the vectors in (5.25) are collinear and we can write:

$$D = \varepsilon_0 E + P \tag{5.26}$$

where D, E and P represent the magnitudes of collinear vectors. So what precisely is electric polarisation and how does it help to suspend an insulating object?

In electrostatics, there are essentially two types of dielectric – polar and non-polar. In a polar material, the molecules that compose it tend to be asymmetric in relation to atomic charge distribution so that the centre of negative charge is displaced from the centre of positive charge thus forming a dipole. A good example is water that is polar because the H_2O molecule is asymmetric. While each hydrogen atom is strongly bonded by sharing electrons covalently with the oxygen atom, the electron 'cloud' of the molecule tends to favour the oxygen nucleus leaving the hydrogen nuclei charge deficient. In her elegantly written book entitled *The Cannon*, Natalie Angier [15] expresses the relationship picturesquely as follows: the molecule 'is best exemplified by the stridently unserious image of Mickey Mouse ... with the head representing oxygen, the ears the two hydrogen atoms covalently linked to it'. Because of the asymmetry 'the ears of the Mickey molecule have a slight positive charge ... the bottom half of the mouse face has a five o'clock shadow of modest negative charge'. In a mass of water the 'chins of one molecule are drawn to the ears of another' so that water molecules cling together just enough to give it its liquid properties.

In the absence of an applied field, as suggested in Figure 5.10b, a polar material with its constituent dipoles randomly orientated remains neutral electrically. On the other hand, when an external field is applied to such a material as suggested in Figure 5.10c, the dipoles become aligned by the field polarising the sample. This alignment process is termed orientation polarisation. It generally results in very large values of relative permittivity – in the case of water $\varepsilon_r = 81$.

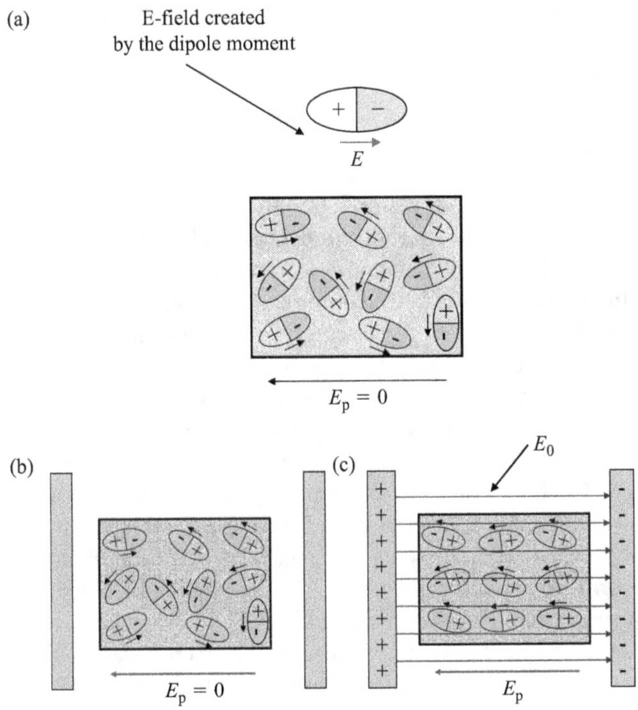

Figure 5.10 (a–c) Influence of an applied electric field E_0 on a polarised dielectric immersed within it

Polar materials are usually liquid, or perhaps gaseous, such as alcohol, hydrochloric acid and ammonia, and consequently are not really of relevance to electrostatic suspension systems. On the other hand, the non-polar materials are solids forming

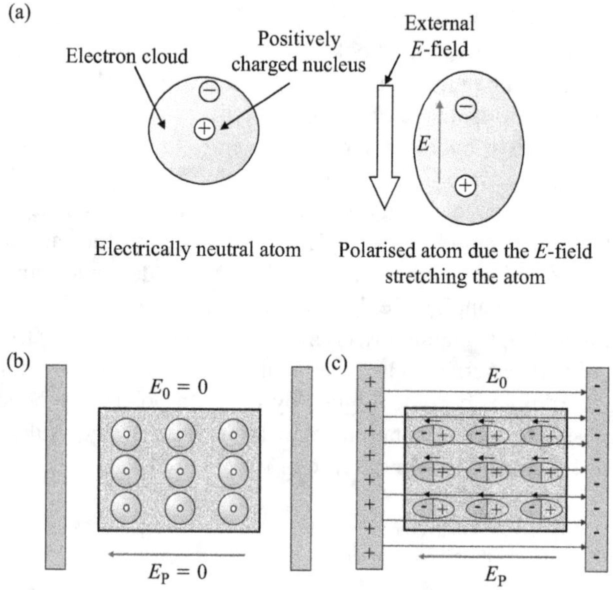

Figure 5.11 (a–c) Polarisation of a non-polar dielectric material by atomic distortion

dielectrics or insulators, which feature strongly in electrical devices, so they are most certainly employable in suspension technologies. For these materials polarisation involves the distortion of neutral atoms within the insulator as suggested in Figure 5.11. In an electric field-free environment the atoms of a non-polar dielectric are electrically neutral with the 'centre of gravity' of the electron cloud co-located with the positive nucleus (Figure 5.11a). However, when the insulator is immersed in a DC electric field (white down arrow in Figure 5.11a) each atom in the material is distorted with the bound electrons being dragged towards the source of the field (positive plate) while the nucleus stays put. A dipole is formed but only when the field is present. Neutrality re-emerges immediately on release of the applied field. This mechanism is emphasised in Figures 5.11b and c, which display a sample of a non-polar insulator located between the electrodes of a parallel plate capacitor. In Figure 5.11b, with no applied electric field ($E_0 = 0$) the sample obviously retains its neutral composition. On the other hand, in Figure 5.11c where E_0 is now finite, and greater than zero, atomic dipoles are formed as the electrons are pulled towards the positive plate, while the positive nuclei remain fixed in position. The insulator, in this case, is distortion polarised, and a macroscopic polarisation electric field E_p is produced between the skin of negative charge that forms on the left face of the sample, and the positively charged skin resident on the right surface, due to the absence of electrons there.

Polarisation, whether by orientation or distortion, permits much higher energy levels to be stored within an electrostatic capacitor because much higher voltage levels than in air, can be sustained without the danger of catastrophic breakdown.

The reason for this is illustrated schematically in Figure 5.12a by examining a parallel plate capacitor with an embedded insulator. The insulator is polarised by the applied field E_0 (due to charge Q_{free} on the electrodes) resulting in a macroscopic charge separation (Q_p) within it (see Figure 5.12b). The outcome is a polarisation field E_p opposed to E_0. Vectorial addition of these two field components within the dielectric dictates that a resultant field E_{total} exists given by

$$E_{\mathrm{total}} = E_0 - E_\mathrm{p} = E_\mathrm{T} \tag{5.27}$$

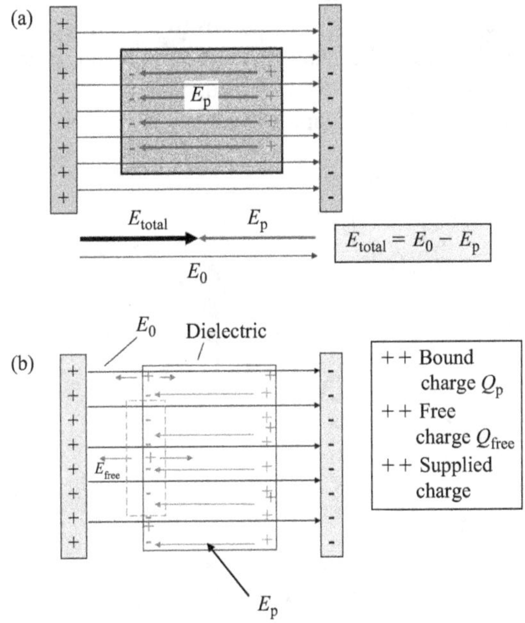

Figure 5.12 (a and b) Field stress reduction in an insulator due to distortion polarisation

But the electric field boundary conditions at the air/insulator interfaces require that normal electric flux density is continuous. Hence we must have:

$$\varepsilon_\mathrm{r}\varepsilon_0 E_\mathrm{T} = \varepsilon_0 E_0 \tag{5.28}$$

which implies that E_T is reduced, by an amount ε_r times, relative to its value in air (E_0) for a given electrode charge Q_{free}. Also on combining (5.27) and (5.28) with (5.26) in the form:

$$D_\mathrm{T} = \varepsilon_0 E_\mathrm{T} + P$$

the relationship between the polarisation field E_p and the polarisation vector P is generated. That is

$$E_\mathrm{p} = -\frac{P}{\varepsilon_0} \tag{5.29}$$

Figure 5.11 makes it clear that the more atoms there are per unit volume of an insulator the more dipoles are formed and hence the greater is E_p for a given applied field level. Equation (5.29) dictates that the polarisation vector per unit volume and the relative permittivity must therefore increase with increasing numbers of atoms per unit volume that equates to increasing material density.

Electrostatic suspension of a float formed from an insulating rather than a conducting material exhibits difficulties that do not arise with metal floats. In the latter case, as illustrated in Figure 5.13, when a metal float is brought into close proximity with a pair of oppositely charged electrodes (Figure 5.13a) free charge on the plate separates to bring negative charge under the positive electrode and positive charge under the negative electrode, the plate as a whole remaining neutral electrically. A strong electrostatic suspension force is directly generated by the powerful charge attraction across the electrode/float gap.

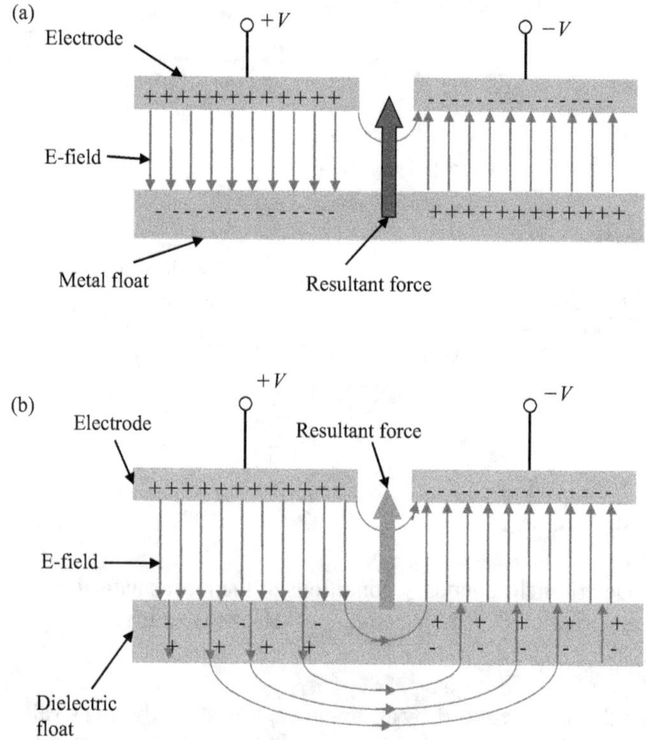

Figure 5.13 Comparison of suspension of (a) metal float and (b) insulating float

When an insulating float is brought into close proximity with a similar electrode geometry (Figure 5.13b), the electrostatic field formed by the electrodes polarises the insulator as suggested in the figure. Once this happens an electrostatic field is set up between the stator electrode charges and the oppositely charged 'skins' on the upper surface of the float. This mechanism is not unlike that pertaining to the metal float.

However, unlike the metal example the electrostatic field penetrates into the dielectric although, as we have seen above, weakened by the polarisation field. This penetrating field produces a downwards, or repulsive, force on the charge-skins on the lower surface of the insulator, thereby reducing the overall force available for suspension. This effect can be reduced by increasing the permittivity of the float material, but this can be taken only so far since high permittivity equates to high density and heavy floats. Alternatively, if the float is slightly conducting [7], then some free charge movement (see Figure 5.12b) controlled by the resistivity level, can be incorporated into the mechanism enhancing the suspension force strength.

5.3 Vacuum applications

The range of applications for electrostatically suspended and propelled devices is growing steadily from bearings [16], gyroscopes [17], MEMS actuators [10], to scientific instruments [18–22]. This latter category has seen major advances in recent years and an example, which is representative of previous discussions in this chapter, is presented below.

Efficacy in the contactless, containerless processing and measurement of materials is an increasingly important aspect of the study of thermo-physical properties of undercooled liquid states, in the investigation of nucleation kinetics, and in the creation of unique metaphysical phases that may lead to novel material properties [20]. Electrostatic suspension techniques provide the means of superheating, undercooling and solidifying metals and alloys in a contamination free (contactless and high vacuum) stable environment. There are many advantages of electrostatic suspension over magnetic levitation in high-vacuum and high-temperature processing environments and these are clearly enunciated in reference 20.

Photographs of a scientific apparatus (termed an electrostatic levitator or ESL) for electrostatic levitation/suspension of material samples in a high-temperature vacuum test and measurement chamber is shown in Figure 5.14. This system is attributable to NASA. A spherical float formed from the material under investigation, which may be conducting or non-conducting, solid or liquid, is suspended in an essentially radial suspension system (see section 5.3.4). The vacuum chamber is shown in Figure 5.14a while the electrode arrangement is clearly visible in Figure 5.14b.

The essential elements of the ESL are shown schematically in Figure 5.15 to re-enforce clarity. Two horizontal disc-shaped electrodes electrically suspend the sample in the vertical direction, with the upper electrode providing lift while the lower electrode provides a stabilising downwards thrust if necessary as in space applications. On earth, downwards thrust is of course supplied by gravity. The sample is maintained in position horizontally by two smaller pairs of electrodes that create a two-dimensional potential well. Sample heating is provided by a high-power deuterium UV-rich arc lamp, the output of which is focused onto the sample. This heating method transfers charge to the sample by photo-electric means, thus enhancing the electrostatic suspension and control forces. Note that with a charged float the electrodes can provide repulsion as well as attraction – a useful feature in space

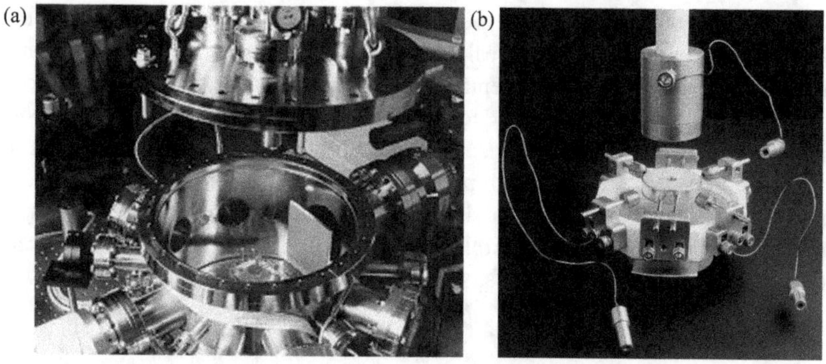

Figure 5.14 *The heart of the ESL is the vacuum chamber (a) containing a pair of electrostatic plates and (b) four electrodes that position the sample being processed. The sample's position is determined from the shadows cast on detectors as two lasers shine at right angles through the vacuum chamber onto the sample [Source: www.esl.msfc.gov]*

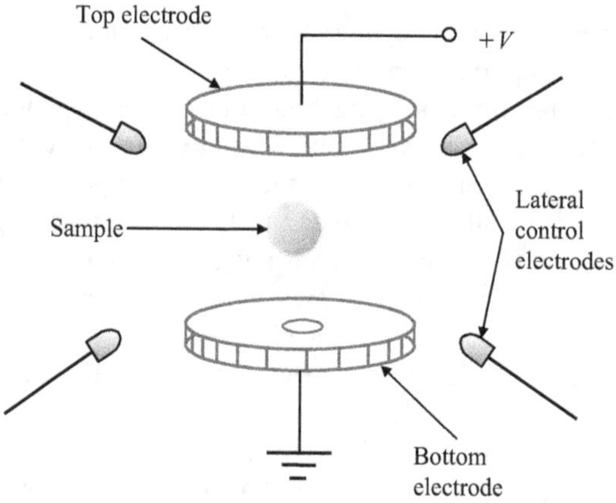

Figure 5.15 *Schematic diagram of the electrode assembly for ESL*

applications. Digital feedback controls the system, using lasers to calculate subtle changes in the electrode charges and recentering the sample so it remains in the path of the heating UV beam – although experiments can also be run without heating. At present the power of the NASA electrostatic levitators is limited, so samples can be no more than 3 mm (0.12 in.) in diameter and weigh 30–40 mg.

5.4 Summary

The Earnshaw rules of suspension/levitation by electromagnetic methods predict that an object suspended by an inverse square law dependent field, such as the electrostatic field, is inherently unstable. Consequently, this kind of suspension cannot be implemented without resorting to sensing and control electronics. The basic equations that encapsulate the nature of the electrostatic suspension forces on metal floats are developed in section 5.1, by applying energy expressions developed in Chapter 3 to a four-electrode example that is typical of the genre.

For multi-electrode systems, the equations governing electrostatic suspension forces are shown to be not unlike those of electrostatic propulsion forces. The contactless transportation of metal discs, which accesses both sets of forces, is discussed in section 5.1.4. The not dissimilar problem of providing stable suspension of the rotor of an axial-drive micromotor is pursued in section 5.2, while at the end of the section the polarisation of dielectrics is examined in detail as a prelude to providing an explanation of the electrostatic suspension of insulators.

It is probably reasonably true to say that electrostatic suspension techniques have been confined to devices already possessing electrostatic functions, such as electrostatic actuators/motors, or to devices incorporating floats that are not susceptible to magnetic fields, or to devices required to operate in environments that are not tolerant of magnetic materials. Fortunately, these applications are growing as indicated in section 5.3, by developments in containerless high-temperature vacuum processing systems.

Frictionless electrostatically suspended actuators will largely feature in products that can be improved by the introduction of MEMS fabricated devices – such as watches, cameras, mobile phones and computer game controllers. From a profligacy of power use viewpoint, as human economies begin to face up to a world without fossil fuels, it seems on the face of it, that these very low power consumption products are unlikely to be fertile targets for efficiency drives. However, there are vast numbers of them now, and they will continue to be so in modern electricity-powered communities of the future. All require power from the grid to function, directly if they have to be recharged, and if they are run on replaceable batteries these need to be manufactured, which takes power. Consequently, improved operational efficiency through the adoption of frictionless techniques could still reap considerable gains even at the micro-scale end of technology development.

5.5 References

[1] Jeon, J.U., Woo, S.J., & Higuchi, T. 'Variable capacitance motors with electrostatic suspension'. *Sensors & Actuators A*. 1999;**75**:289–297

[2] Jeon, J.U., & Higuchi, T. 'Induction motors with electrostatic suspension'. *Journal of Electrostatics*. 1998;**45**(2):157–173

[3] Niino, T., Eto, K., & Higuchi, T. 'Basic study of electrostatic levitation system in high vacuum condition'. *IEEE Conference on Industry Applications*; Rome, 2000, pp. 682–686

[4] *Amperes: 3D magnetic field simulation software*. Available from www.integratedsoft.com/products/amperes/default.aspx

[5] *ANSYS Maxwell*. Available from www.ansoft.com

[6] Woo, S.J., Jeon, J.U., Higuchi, T., & Jin, J. 'Electrostatic force analysis of electrostatic levitation system'. *SICE*; Sapporo, Japan, 1995, pp. 1347–1352

[7] Jin, J., & Higuchi, T. 'Direct electrostatic levitation and propulsion'. *IEEE Transactions*. 1997;**IE-44**(2):234–239

[8] Jin, J., Higuchi, T., & Kanemoto, M. 'Electrostatic levitator for hard disc media'. *IEEE Transactions*. 1995;**IE-42**(5):467–473

[9] Bollee, B. 'Electrostatic motors'. *Phillips Technical Review*. 1969;**30**: 178–194

[10] Bart, S.F., Lober, T.A., Howe, R.T., Lang, J.H., & Schlecht, M.F. 'Design considerations for micromachined electric actuators'. *Sensors & Actuators A*. 1988;**14**:269–292

[11] Trimmer, W., & Jebens, R. 'An operational harmonic electrostatic motor'. *Proceedings of IEEE*; Microelectromechanical Systems Workshop, Salt Lake City, USA, 1989, pp. 13–16

[12] Jacobsen, S.C., Price, R.H., Wood, J.E., Rytting, T.H., & Rafaelof, M. 'The wobble motor: an electrostatic, planetary armature, microactuator'. *Proceedings of IEEE*; Microelectromechanical Systems Workshop, Salt Lake City, USA, 1989, pp. 17–24

[13] Sangster, A.J., Reuben, R.L., Samper, V., & Shea, K. 'An electrostatic harmonic microactuator for arterial plaque removal'. *Institute of Physics, Journal of Micromechanics and Microengineering*. 1995;**5**:297–303

[14] Kraus, J.D. *Electromagnetics*. London: McGraw-Hill Book Co.; 1984

[15] Angier, N. *The Cannon*. New York, NY: Houghton Mifflin Co.; 2007

[16] Han, F., Wu, Q., & Gao, Z. 'Initial levitation of an electrostatic bearing system without bias'. *Sensors & Actuators A*. 2006;**130–131**:513–522

[17] Knoebel, H.W. 'The electric vacuum gyro'. *Control Engineering*. 1964; **11**(2):70–73

[18] Witt, D.C. 'A spinning rotor vacuum gauge with electrostatic suspension'. *Vacuum*. 1995;**46**(4):397–405

[19] Rhim, W.K., Collender, M., Hyson, M.T., Simms, W.T., & Elleman, D.D. 'Development of an electrostatic positioner for space material processing'. *Review of Scientific Instruments*. 1985;**56**:307–317

[20] Rhim, W.K., Chung, S.K., Barber, D., Man, K.F., Rubison, A., & Spjut, R.E. 'An electrostatic levitator for high temperature containerless materials processing in 1-g'. *Review of Scientific Instruments*. 1993;**64**(10):2961

[21] Rulison, A.J., Watkins, J.L., & Zambrano, B. 'Electrostatic containerless processing system'. *Review of Scientific Instruments*. 1997;**68**(7):2856–2863

[22] Rhim, W.K., & Paradis, P.-F. 'Laser-induced rotation of a levitated sample in vacuum'. *Review of Scientific Instruments*. 1999;**70**(12):4652–4655

Chapter 6

Electrical engineering perspective on superconductors

> [science] is simply common sense at its best – rigidly accurate in observation, and merciless to fallacy in logic.
>
> *Thomas Huxley*

6.1 Superconduction and electron waves

When the Dutch scientist, H. Kamerlingh Onnes [1853–1926] observed *suprageleider* (superconductivity to the English speaking world) in mercury in 1911 by immersing it in liquid helium, he was aware of its potential for advancing the influence and reach of magnetic technology. Superconductors are not just better conductors of electricity than well-known metals such as copper, silver and aluminium, but they represent a completely different phenomenon in electrical science, as is suggested by the resistance versus temperature graphs in Figure 5.1, originally due to Onnes.

In general, most conventionally conducting materials display an increasing resistance with temperature, because fixed ions within the material become more agitated as it becomes hotter. When viewed as particles, as is normally the case in electrical engineering, 'free electrons' in the hot solid experience a much more disruptive and tortuous path through it, because of the interfering vibrating ions, and hence the material is deemed to exhibit high resistance. But even at absolute zero (0 K), such materials possess some residual resistance because the electrons as particles are still impeded by the lattice of atoms (Figure 6.1 – curve with triangles). For a superconductor the behaviour at low temperature is quite different, with zero resistance or perfect conductivity being possible (curve with squares). There is a critical temperature (T_c) above which the superconductor acts normally, exhibiting high resistivity (~4 K in Figure 6.1). Below this temperature it switches to a zero resistance state. The electrons are unimpeded as they move through the superconductor even although it still possesses a lattice of fixed ions. So how do the electrons behave so differently, in superconductors as compared to conventional conductors, below T_c?

The answer to the above questions, although still incomplete, lies in quantum mechanics and there are many general science and specialist books available that attempt an exploration of superconductivity from this viewpoint [1–3]. However, electrical engineers live and operate in the macroscopic world and seldom, if ever,

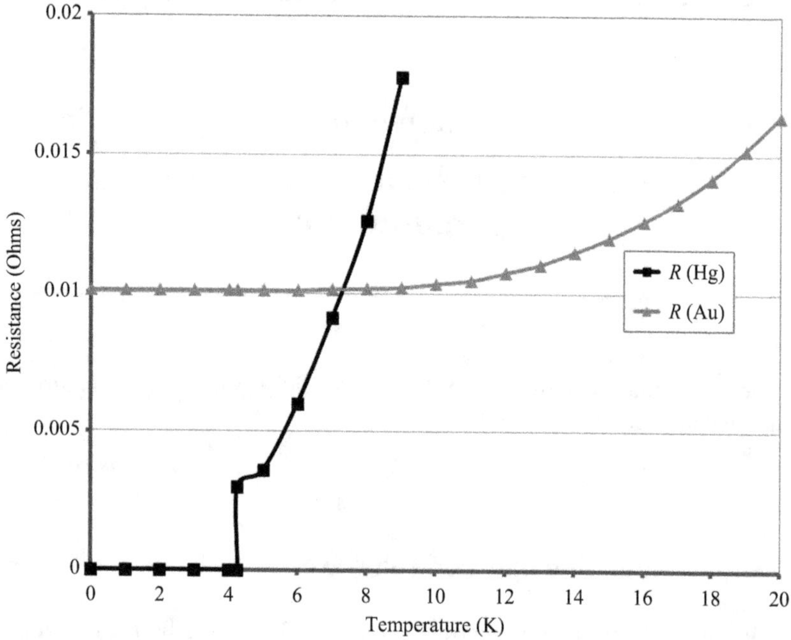

Figure 6.1 Comparison of the temperature behaviour of a conventional and a superconductor: gold = triangular symbols; mercury = square symbols [data from Onnes]

enter the rarefied microscopic realm of individual atoms and quantum behaviour. So quantum explanations of the superconducting phenomenon, which obviously has significant implications for modern engineering practice, are both elusive and singularly 'missing the target'. Yet engineers aiming to incorporate superconductivity into practical devices and systems need a good understanding of the phenomenon; one which addresses macroscopic interactions and is robust enough to encourage and support innovation. Section 6.1 attempts to provide such a treatment.

That superconductivity is a distinctly odd phenomenon in electrical science manifests itself by the fact that classical electromagnetism [4–9] and classical electrodynamics [5] are no longer reliable analysis tools when applied to 'simple' electrical circuits containing a superconducting component.

6.1.1 Classical interpretation

Let us take a look at an electrical engineering example that graphically demonstrates how a quantum effect (superconductivity) can impinge unexpectedly on the classical macroscopic world familiar to engineers. It demonstrates that a classical treatment of this apparently simple electrical circuit results in an incomplete exposition. Quantum mechanics [10–12] is required to complete the picture. The electrical engineering

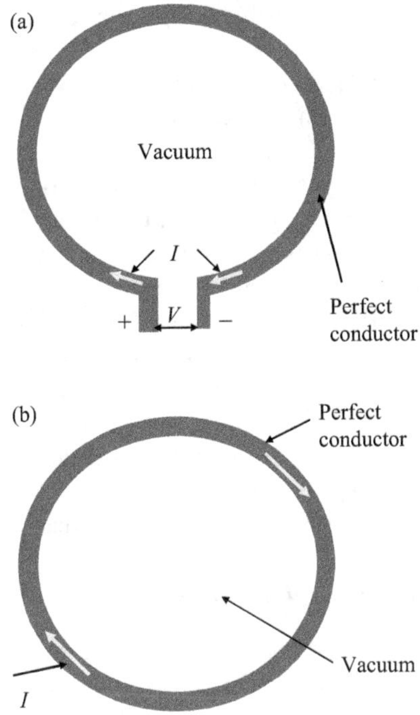

Figure 6.2 Superconducting loop in vacuum (a) open, (b) closed

conundrum can easily be stated by considering the behaviour of electrons (current) in a superconducting loop [12] as indicated in Figure 6.2.

In Figure 6.2a, the superconducting loop represents a simple one-turn inductor, which will be familiar to most second year undergraduates, pursuing an electrical engineering degree course in Britain. The inductance L (Henries) is dependent on the loop dimensions. If the current is initially zero and rises to a final value I ($0 < i < I$) in time T, then familiar circuit relations yield

$$V = \frac{d\Phi}{dt} = L\frac{di}{dt} \tag{6.1}$$

or at time T

$$\Phi = LI = \int_0^T V\,dt \tag{6.2}$$

Also, at time T the energy stored in the loop (W) is

$$W = \int_0^T Vi\,dt = \frac{1}{2}LI^2 \quad \text{(Joules)} \tag{6.3}$$

In classical electromagnetism [6–9], we would infer that this energy is stored in the magnetic field. In electrodynamics, on the other hand, each moving electron in the loop couples with every other electron through 'action at a distance' so storing energy in the interaction process.

Now, if the perfectly conducting and isolated loop is closed as suggested in Figure 6.1b with the current I continuing to flow, an energy W can be stored in the loop in perpetuity since there is nothing to impede the flow of the moving electrons. Scientists have shown that it is possible to close such a loop without quenching the current. A *persistent* current is formed. But on the basis of a classical electromagnetic (CEM) analysis of this macroscopic problem, the expectation arising from the Faraday law (6.1) would be that if $V = 0$ for the closed loop, then the rate of change of flux must be zero $\left(\frac{d\Phi}{dt} = 0\right)$. In contradiction to experiment, CEM seems to predict that it should not be possible to develop a current in, or a flux through, the perfectly conducting ring. This is because CEM views current as a swarm-like continuous movement of charge through the conductor, thus failing to accommodate the now well-known Meissner effect [11]. The result is that energy storage in the macroscopic scale, perfectly conducting ring, escapes explanation by CEM. The answer to this dilemma can be found in cumulative quantum electrodynamics (QED) [12].

6.1.2 *Cumulative quantum electrodynamics*

Measurements [12] on current carrying superconducting loops of varying dimensions, reveals that flux is present and quantised and that (6.2) should more correctly read

$$\Phi = \int V \, dt = n\Phi_0 \tag{6.4}$$

This equation suggests that at the microscopic level the electric potential (V) within the loop material is not zero. Experiments, accurate to one part in 10^9, indicate that the flux quantum $\Phi_0 = 2.06783461 \times 10^{-15}$ V-s. The explanation for these behaviours can be deduced from QED that focuses on more fundamental physical parameters such as energy, momentum, frequency and phase. Einstein has provided the relation for energy (W)

$$W = \hbar\omega \quad \text{(Joules)} \tag{6.5}$$

while de Broglie has determined that momentum

$$\vec{p} = \hbar\vec{k} \quad (\text{kg m/s}) \tag{6.6}$$

The constant of proportionality $\hbar$ is just Plank's constant divided by 2π. These equations dictate that at the quantum (atomic) level energy is synonymous with frequency while momentum is synonymous with phase. The quantity $\vec{k}$ is defined as the propagation vector of a wave in rad/m. When applied to an electron, this can only mean that the electron has wave characteristics, so that a vast number of interacting and coupled electrons in the superconducting ring form a coherent wave such that the cumulative phase around the loop is

$$\phi = \oint_{\text{loop}} \vec{k} \cdot d\vec{\ell} \quad \text{(radians)} \tag{6.7}$$

Measurements also reveal that changing n to $n+1$ or $n-1$ in a fixed length ring is very difficult – presumably because electrons come in essentially one form, with a tightly specified energy level and hence a limited frequency range. The superconducting loop is in essence a ring resonator for the cumulative electron wave. It dictates that the loop-phase is quantised such that $\varphi = 2n\pi$. The reason for the quantum behaviour is thus apparent – the wave nature of the electron ensemble. Given that accumulated phase equates to the time-integrated frequency difference between two points in the loop and that frequency is synonymous with energy, it can be shown [12] that

$$\phi = \frac{q_0}{\hbar}\int V\,\mathrm{d}t \tag{6.8}$$

The non-zero voltage is the evidence of an electron wave. Hence combining (6.8) and (6.2) we get

$$\phi = \frac{q_0}{\hbar}\phi = \frac{q_0}{\hbar}n\Phi_0 \tag{6.9}$$

So magnetic flux can be equated to phase accumulation of the electron-wave function, which is usually represented by the magnetic vector potential $\vec{A}$ and by analogy with (6.7) this leads to the following equation:

$$\Phi = \oint_{\text{loop}} \vec{A}\cdot\mathrm{d}\vec{\ell} \tag{6.10}$$

Furthermore, mutual coupling experiments using loops of the form suggested in Figure 6.1 indicate that the interaction between the loops associated with the action at a distance of the electrons in each loop, can for any loop size, shape, orientation and separation, be represented by the vector potential $\vec{A}$. It can be calculated by

$$\vec{A} = \frac{\mu_0}{4\pi}\oint_{\text{loop}} \frac{\vec{I}}{r}\mathrm{d}\ell \tag{6.11}$$

where $\vec{I}$ is the current in the loop and μ_0 is the permeability of vacuum. Equations (6.7), (6.9) and (6.10), and a generalised form of (6.11), are the fundamental relations governing quantum electrodynamics in *cumulative form* [12], which recognises that electrons have wave characteristics and do not exist as totally isolated, 'point-form', particles. The German physicist Fritz London [1900–1954] coined the phrase 'macroscopic quantum phenomenon' to describe this aspect of superconductivity [13].

6.2 Resonance and coherence

While it is clear from the above that a macroscopic quantum phenomenon occurs in a superconducting ring, it is still not obvious why the electrons become wave-like and choose to couple and cohere in a suitable material when it is operated below a critically low temperature.

6.2.1 Ring resonator

From an engineering perspective, this question can be illuminated quite satisfactorily by resorting to an analogous resonant ring problem that is not unfamiliar to electrical engineers who have studied radio waves – a model that may, or may not, be appreciated by quantum physicists! For simplicity, the rectangular waveguide ring (Figure 6.3) is assumed to be formed from infinitesimally thin highly conducting walls, and to trap a vacuum in the interior space. In Figure 6.3a, the ring of average length L metres, which is presumed to be much larger than guide width a, is fed at one port by a pulsed source of frequency f_0. The second port is terminated in a matched load. In Figure 6.3b, the ring is closed to form a ring resonator.

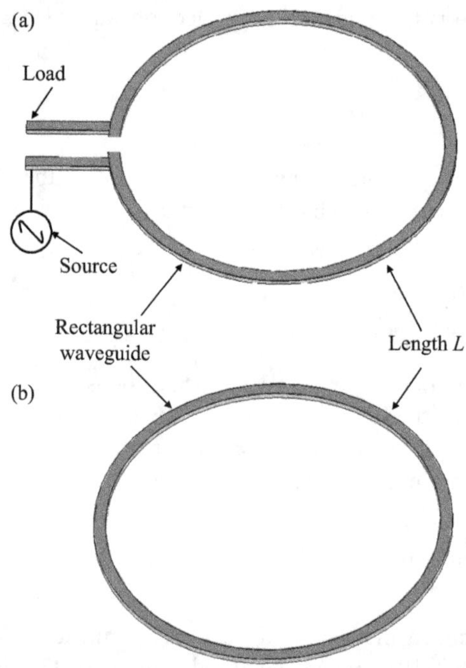

Figure 6.3 (a) Waveguide loop with matched load (b) waveguide ring resonator. In both cases the waveguide has width a, height b, length L with perfectly conducting walls and evacuated interior

This waveguide geometry, is described mathematically in Chapter 2 in accordance with CEM theory [6–9]. The analysis commences with Maxwell's equations and leads to the formation of a second-order differential equation, in terms of the electric field vector or the magnetic field vector (**E** or **H**), which is applicable to the interior region. When these are solved for the rectangular guide, the axial H-field solution for the TE_{mn} mode is

$$H_{zmn} = C \cos \frac{m\pi x}{a} \cos \frac{n\pi y}{b} \exp -\bar{\gamma} z \exp j\omega t \tag{6.12}$$

The solution is dependent on the following relationship between frequency and guide dimensions being established (see Chapter 2):

$$\bar{\gamma}^2 = \bar{\gamma}_{mn}^2 = \left(\frac{m\pi}{a}\right)^2 + \left(\frac{n\pi}{b}\right)^2 - k^2 = k_{cmn}^2 - k^2 \tag{6.13}$$

Here $k = \omega/c$, where k is the free-space propagation coefficient, k_{cmn} is the cut-off coefficient, while γ_{mn} is the waveguide propagation coefficient for the mnth solution or mode. The other field components of the mode are derived from H_z using relations derived from the Maxwell equations (see Chapter 2).

Equation (6.12) tells us that in the terminated waveguide example (Figure 6.3a), the field solutions are modal solutions associated with a double infinity of transverse standing wave forms. These are termed the TE_{mn} modes. The rectangular waveguide also supports a double infinity of TM_{mn} modes. While the system exhibits modal solutions, it is important to note that the frequency (ω) of any given solution is solely dictated and controlled by the source. In other words, the entire system is linear.

6.2.2 *Pulses to waves*

Let us consider exciting the waveguide, as in Figure 6.3a, from a pulsed free-running source so that the electromagnetic wave pulses are incoherent (in arbitrary phase). If the source coupler is designed to excite only the TE_{10} mode (the most common arrangement), a very long pulse, of duration much greater than the propagation delay, will be represented in the guide by a TE_{10} mode at the carrier frequency, the sidebands will be few and very close to the carrier and the guide will seem to support a single continuous wave (CW) signal whose frequency is wholly controlled by the source. Nevertheless, the trapped waves associated with each pulse of power from the source will remain phase incoherent. On the other hand, if the waveguide extremities are now connected together to form a low loss (high Q) ring resonator (Figure 6.3b), and L is an integer number (p say) of wavelengths at the carrier frequency, this frequency will be locked into the resonator. Now the ring resonator is not unlike the superconducting ring, except that the 'wave action' takes place within the vacuum space rather than within the perfectly conducting material. The electromagnetic solution for the ring proceeds as described in Chapter 2, but with the additional boundary condition that the phase around the ring for both clockwise and anticlockwise waves must be an integer number (p) of cycles. The resultant standing wave solution has the form

$$H_{zmn} = C \cos \frac{m\pi x}{a} \cos \frac{n\pi y}{b} \cos \beta_{mn} z \exp j\omega t \tag{6.14}$$

where

$$\beta_{mn} = \frac{2p\pi}{L} \tag{6.15}$$

In the absence of loss, a non-linear 'pumping process' will develop strong 'ringing' in the resonator from the incoherently phased pulses. But significantly, the ringing circular cavity could, in principle, become a source of a coherent radiation signal.

The original incoherent signal has become coherent due to the filtering action of the resonator.

For very short pulses, the transition to coherency is more difficult to rationalise. A short pulse, of time duration much less than the propagation delay through the waveguide, will be represented by a spectrum of sidebands on either side of the carrier frequency, all propagating in the TE_{10} mode, as the pulse passes through the guide. If the fundamental pulse frequency is sufficiently remote from cut-off, so that the guide is essentially non-dispersive for the broad range of this frequency spectrum, then the pulse will traverse the guide largely unchanged. Now, when the guide is closed to form the ring resonator, the wideband pulses can no longer exist in the high-Q structure. The pulses merely provide pumping energy to sustain the resonance at the carrier frequency. Once again, a coherent wave has been extracted from incoherent pulses.

A critical aspect of this process, which is relevant to the superconducting ring, is that the formation of a coherent resonance within the high-Q ring waveguide is dependent on the guide being highly stable dimensionally and materially so that the ring resonant frequency is fixed. If the ring waveguide were to be exposed to a cycling and erratic movement in temperature resulting in dimensional fluctuations, then the pulses entering the system would not induce a coherent resonant mode. The incoherent pulses would simply travel around the ring in much the same way as they propagated around the open waveguide (Figure 6.3a).

6.3 Perfect conductivity

With an appreciation of modal behaviour in an electromagnetic waveguide resonator, we are now in a position to provide a plausible *engineering* explanation of superconduction without getting 'bogged down' in quantum mechanics. At very low temperature where the ionic lattice in a material such as mercury is devoid of vibrational energy, the lattice becomes a filter (like waveguide – particularly periodic waveguide) whose passband is aligned to frequencies associated with electron waves. Like the waveguide ring resonator the incoherent electron pulses provide pumping to a natural coherent mode (macroscopic quantum phenomenon) of the superconducting ring that eventually provides the whole mechanism of conduction in the ring (Figure 6.1b). As with the waveguide resonator the stability of this mode is disturbed and quenched if the ionic lattice begins to vibrate by applying heat and raising its temperature to the critical value.

The key step towards comprehending the superconduction of electrons, in quantum terms, was the observation that one electron (of an electron pair) passing through the material could seemingly ease the progress, of the second electron, through the ionic lattice in its path by helpfully deforming it, so shifting ionic obstacles out of the paired electron's way. This coupled electron phenomenon was made more explicable by the realisation that at superconducting temperatures electron–lattice interaction creates a weak but attractive force of sufficient strength to encourage electrons to pair. The mechanism is termed the Cooper pair. Leon Cooper [b. 1930] in cooperation with fellow Nobel Prize winners John Bardeen [1908–1991], and J. Robert Schrieffer [b. 1931], built on this concept to formulate an electron-wavefunction (termed the Bardeen–Cooper–Schrieffer (BCS) wave function) that accommodates the idea of fixed phase and a coherent state – also observed in electromagnetic waveguides as we have seen.

In the early 1930s measurements on superconductors due to Walter Meissner [1882–1974] and Robert Ochsenfeld [1901–1993] seemed to suggest that any magnetic field in the superconductor's vicinity was completely excluded from the interior. However, later theoretical studies initially due to the London brothers showed that this was not the case. This can be demonstrated by considering the origins of (6.11). This cumulative electron dynamics equation [12] is a solution of the second-order differential equation (cf. (2.14) and (2.15)):

$$\nabla^2 \mathbf{A} = -\mu_0 \mathbf{J} \tag{6.16}$$

But for N elementary charges per unit volume, current density $\mathbf{J} = q_0 N \mathbf{v}$, where q_0 is the electron charge, and $\mathbf{v}$ is its velocity. In electrodynamics, electron momentum is equal to $q_0\mathbf{A}$ for a macroscopic electron wave, therefore

$$\mathbf{J} = \frac{q_0^2 N}{m} \mathbf{A} \tag{6.17}$$

Hence we deduce that

$$\nabla^2 \mathbf{A} = \frac{\mu_0 q_0^2 N}{m} \mathbf{A} \tag{6.18}$$

This differential equation has a solution of the form

$$A = A_0 \exp \frac{-x}{\lambda} \tag{6.19}$$

with $\lambda^2 = \frac{m}{\mu_0 q_0^2 N}$ and x assumed normal to the conductor surface. The form of (6.19) represents a typical decay function, indicating that magnetic field (A) is not totally excluded from the interior of the superconductor. It penetrates perceptibly, then decays very rapidly with distance. This is governed by λ that is variously termed the screening distance, the penetration depth and the skin depth. For typical superconductors, it has a value in the range of 50 nm.

While in conventional conductors electron behaviour can be simulated with acceptable accuracy by treating electrons as negatively charged particles, which create resistance and heat by 'collisions' with fixed ions, in superconductors this model breaks down as demonstrated above. In a solid material forced into the superconducting state by cooling, cumulative quantum dynamics suggest that the electron waves cohere and the macroscopic 'cumulative wave' carries current with no mechanism for ionic collisions. The super-cooled material has become a perfect conductor exhibiting no electrical resistance. Unfortunately, there is a catch, as we have seen. The operating temperature of the superconductor has to be close to absolute zero (−273°C) if it is a pure element. Nevertheless, since Onnes' experiments with mercury, 31 elements have been discovered to exhibit superconductivity at ambient pressure (dark shading in the periodic table of Figure 6.4) while another 22 elements have been

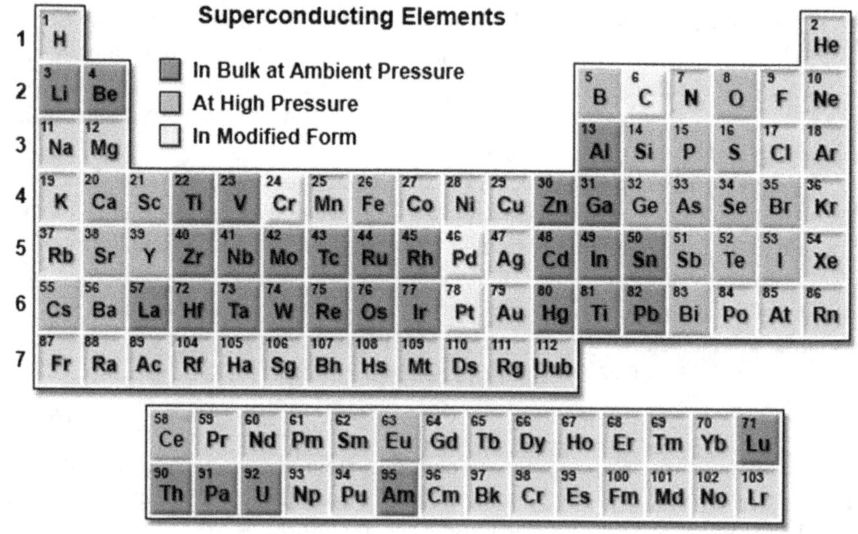

Figure 6.4 Periodic table. Elements that superconduct at ambient pressure are darkly shaded, while those that do so at high pressure are shown lightly shaded

added to the list, which are superconducting at high pressure (lightly shaded squares in Figure 6.4). Four others (white) have been identified, which superconduct but only in a non-pure form.

6.3.1 Alloys

The property that is used to distinguish superconducting materials is the Meissner effect, which encapsulates their ability to counteract or negate magnetic flux in their interior. This is demonstrated schematically in Figure 6.5, for a cylindrically shaped conductor located in a transverse magnetic field. In Figure 6.5a, with the operating temperature above the critical threshold, the magnetic field passes through the interior of the cylinder just as it would for any conventional non-magnetic metal. On the other hand, below T_c, superconduction dictates that currents are instantaneously created (Lenz Law or Meissner effect) to generate a magnetic field that exactly counteracts the applied field (Figure 6.5b). This tendency of a material to diminish the flux within it is termed diamagnetism, and is quite distinct from paramagnetism prominently exhibited by iron or ferrite materials as described in Chapter 3. There we have noted that for any material we can write:

$$\mathbf{B} = \mu_0[\mathbf{H} + \mathbf{M}] \tag{6.20}$$

where **B** is the internal magnetic flux density, **H** is an external applied field and **M** is the magnetic dipole moment per unit volume representing the atomic response on the element or compound. **M** is often expressed as $\mathbf{M} = \chi_m \mathbf{H}$, where χ_m is referred to as the magnetic susceptibility. In a superconductor below T_c, **B** must be precisely zero and therefore $\chi_m = -1$. This negative susceptibility is the reason for superconductors

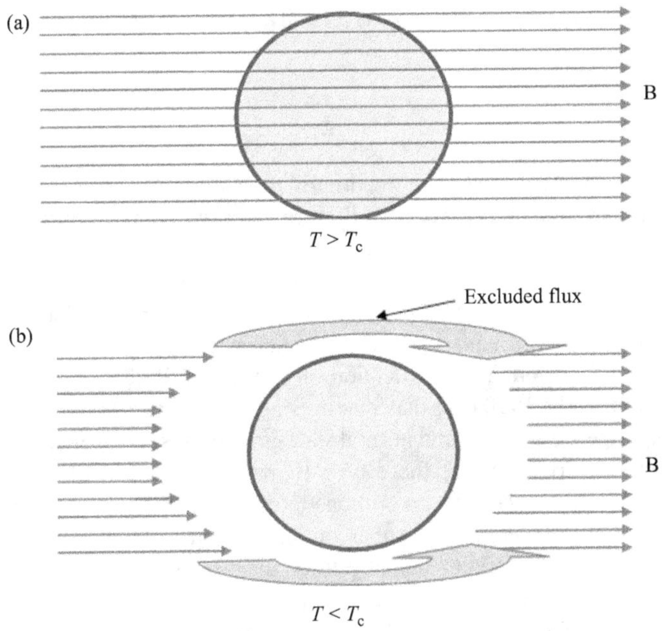

Figure 6.5 Illustration of Meissner effect. (a) Type I superconductor above critical temperature, (b) superconductor below critical temperature

to be classed as diamagnetic. However, for most diamagnetic materials the susceptibility, while negative, is extremely small. For example, copper has a susceptibility of -0.98×10^{-5}. For the superconductor, $\chi_m = -1$ because an infinitesimally thin electron-wave current at its surface maintains exclusion when $T < T_c$. If this screening current has density **J** A/m, then application of the Lorentz law determines that there will be a force ($\mathbf{F}_n$) normal to the surface given by

$$\mathbf{F}_n = \mathbf{J} \times \mathbf{B} \tag{6.21}$$

This result has considerable bearing on levitation with superconductors as we shall see.

Exclusion costs energy to the electron wave, but so does a thin skin of normal material at the periphery that develops heat. More particularly, flux exclusion is undermined or modified in the interests of energy minimisation, as dictated by the first law of thermodynamics. A similar energy balance occurs in electromagnetic waveguides and transmission lines with lossy conductors, and optimum energy transfer leads to a tightly defined skin depth. In type I superconductors the exclusion depth is very small and these materials (mercury, lead, tin etc.) favour a minimal interface between regions in the superconducting state and those in the normal state.

As we have already seen superconductors of this type only exist at or below a certain critical temperature T_c that depends on the element. However, the superconducting state can also be quenched by a strong magnetic field (B_c). If the current density in the superconductor is low, B_c is related to temperature through the following relation:

$$B_c = B_{c0}\left(1 - \frac{T^2}{T_c^2}\right) \tag{6.22}$$

The quenching magnetic field is a maximum at absolute zero (B_{c0}) and decreases rapidly as the temperature approaches the critical value. This is important to the possible use of superconductors in transmission lines, where the transport or signal current can be a source of an internal, potentially quenching, magnetic field.

In alloys or compounds, reduced in temperature to the superconducting state, the energy balance associated with magnetic flux exclusion tilts much more in favour of allowing flux penetration. The implication of this, as realised by Alexei Abrikosov [b. 1928] in the early 1950s, is that maximising the extent of interfaces between regions in the superconducting and normal states, minimises the energy requirements of the system. It turns out that these type II materials are full of interfaces (see Figure 6.6). Abrikosov was able to demonstrate that for materials of this type the magnetic field penetration into the superconductor occurs within tubes of 'normal' conductor arranged in a regular (usually triangular) lattice. The superconductor is shielded from the flux tubes by current vortices as indicated in Figure 6.6, with each tube containing precisely a quantum of magnetic flux. This mechanism explains why many alloy superconductors seem to disobey the Meissner effect by allowing significant flux penetration.

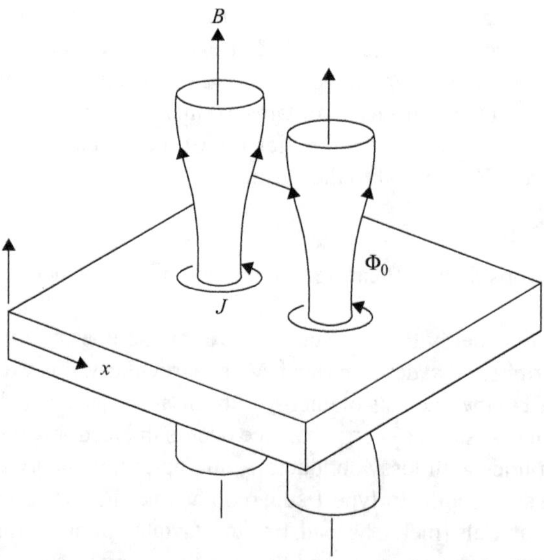

Figure 6.6 Schematic illustration of vortices in type II superconducting alloy in thin film form. Flared tubes represent magnetic vortices, and the rings in superconducting layer represent shielding current loops

Table 6.1 Critical conditions for some type II materials

Material	**Critical temperature (T_c, K)**	**Transition magnetic field (B_{c2}, T)**	**Current density limit (J_c, A/cm^2 (10^5))**
Nb-Ti	9.6	8	2
Nb_3Sn	18	23	3
$YBa_2Cu_3O_7$ (bulk)	95	18	0.3
$YBa_2Cu_3O_7$ (thin film)	95	35–200	600

For these type II materials, the critical magnetic field H_c exhibits two levels (H_{c1} and H_{c2}) that are crucial to the nature of the superconduction states. Below H_{c1} the material will behave like a type I metal enforcing complete field exclusion, whereas if the applied field is above H_{c1} but below H_{c2} type II behaviour takes over with flux penetration permitted within the flux tubes and excluded elsewhere, as suggested in Figure 6.6. T_c and B_{c2} are tabulated for some typical examples in Table 6.1. With critical temperatures above liquid nitrogen (77 K) and substantial levels of tolerance to applied magnetic fields and hence transport currents, the yttrium-barium-copper oxides are eminently practical materials for low loss electrical power or signal transmission. The form of the molecule for this type of material is presented in a clear and instructive graphic in Figure 6.7.

6.3.2 Flux pinning

Type II superconductors have been applied with some success as high current solenoids in electromagnets. Huge magnetic flux densities of over 10 T, with negligible dissipation in the coil have been reported [1]. A conventional copper solenoid in a water-cooled electromagnet of similar flux capacity would probably incur a steady heat dissipation due to resistive losses of the order of 2 MW. In a future energy-strapped world where efficiency will become a paramount engineering goal such savings in ubiquitous electromagnets, from micromotors to levitated trains, will be strenuously pursued.

However, there is a penalty, and it lies in the enhanced manufacturing difficulties presented by superconducting materials in the type II category. When incorporated into an electromagnet, such material not only has to exhibit a critical field level, which is higher than the field to be generated, but must while immersed in this field carry substantial currents without resistive loss. The criterion can be met by brittle materials such as Nb_3Sn that has $H_{c2} \sim 2 \times 10^7$ A/m, while materials with less challenging mechanical properties can achieve 10^7 A/m. For the emerging copper oxide based superconductors H_{c2} levels can be even higher by a factor of 10 or more.

At high field and current levels, as are manifested in electromagnets, power dissipation continues to be troublesome even in superconducting coils. The origin of this dissipation is not hard to discern emanating as it does from the Lorentz law (6.21), applied to the transmission current passing through the type II superconductor and

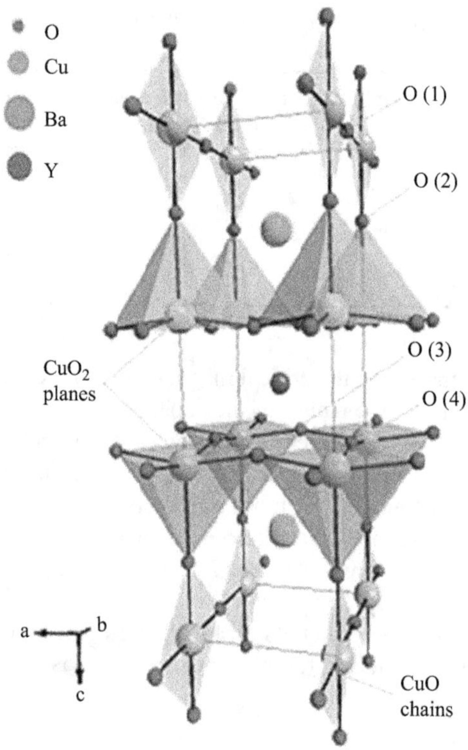

Figure 6.7 Atomic structure of a Perovskite Crystal: namely, Yttrium-Barium-Copper Oxide (from Wikipedia)

the magnetic field threading through a fluxoid or vortex. Hence the force per unit length of a fluxoid (f_v) becomes

$$f_v = \mathbf{J_m} \times \mathbf{a}_y \Phi_0 \tag{6.23}$$

Here $\mathbf{J_m}$ (=$\mathbf{M} \times \mathbf{n}$) is the vortex current screening the fluxoid field from the body of the superconductor, $\mathbf{n}$ is a radially directed unit vector and Φ_0 is the flux quantum. $\mathbf{a}_y$ is a y-directed unit vector. This equilibrium set up is occasioned by the presence of leakage magnetic field permeating the superconductor that is excluded or banished except within the fluxoids. The force, in this case, merely represents outward pressure on the bulk superconductor, and the situation could persist indefinitely. However, when the superconductor supports an externally applied and x-directed transmission current of density $\mathbf{J_x}$, a rather different phenomenon arises. Now Lorentz interaction between $\mathbf{J_x}$ and the fluxoid flux Φ_0 yields

$$f_z^T = \mathbf{J_x} \times \mathbf{a}_y \Phi_0 \tag{6.24}$$

The force f_z^T is clearly transversly directed pushing the fluxoid towards the periphery of the superconductor. If it does so at velocity $\mathbf{v}$ (say) then the resultant

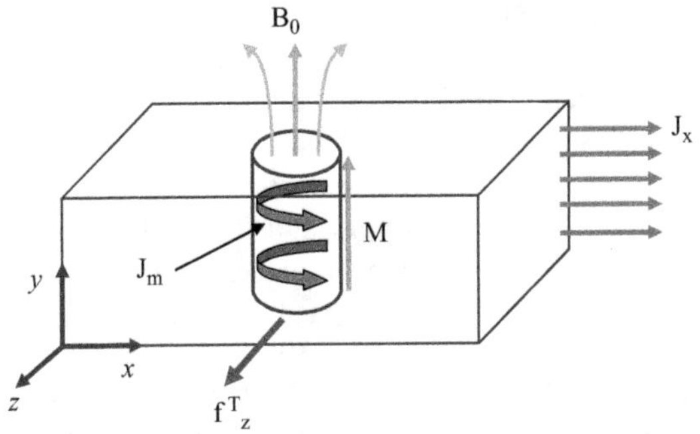

Figure 6.8 *Force on a fluxoid in a type II superconductor due to current transmission in the x-direction*

changing magnetic field, according to Faraday's law, will induce an electric field $\mathbf{E}_i$ given by

$$\mathbf{E}_i = \mathbf{B}_0 \times \mathbf{v} \tag{6.25}$$

This electric field is x-directed, which means that there is a voltage gradient in the direction of the imposed current $\mathbf{J}_x$. The consequence is a mechanism for power loss that manifests itself as resistance to the applied current. This is a seriously debilitating result for type II superconductor applications unless this fluxoid drift can be counteracted.

Stabilising mechanisms that prevent vortex movement due to Lorentz forces are termed *pinning* forces since they, in effect, 'pin' the fluxoids to preferred locations within the superconductor [14]. The fixative generally takes the form of spatial inhomogeneities in the material intentionally created by clever processing methods (grain boundaries, voids etc.), thus providing favoured positions for the vortices. Flux pinning is specifically associated with microstructures in the type II superconductors that tend to inhibit the fluxoids from drifting laterally due to Lorentz forces [3]. The technique is well established and if the pinning is robust fluxoid motion can largely be eliminated, in which case the type II superconductor behaves very much as a perfect conductor. Unfortunately, this type of pinning is not inviolable if overly strong persistent currents are applied. In such circumstances, thermal agitation causes the vortices to skip from one pinning site to another at a discernable but much lower rate than in an unpinned conductor. A mechanism for power absorption and resistive loss, as described earlier, again manifests itself. This regime is referred to as *flux flow* and clearly has to be avoided in practical applications by the control of current levels. Some typical current density limits for a few practical materials are listed in Table 6.1 (column 4).

6.3.3 Superconducting permanent magnets

Superconductors, in addition to reacting to a magnetic field in diamagnetic fashion with a susceptibility of -1, can also be made to trap a persistent self-magnetic field so forming a permanent magnet [3]. To do this, a bulk rectangular or cylindrical specimen of RE-Ba-Cu-O (say, where RE equates to a rare earth element such as yttrium (Y) or samarium (Sm)) is cooled while immersed in a strong magnetic field of from 1 to 20 T. When the sample is removed from the applied field it is observed to contain *persistent*, rather than reactive, circulating currents (vortices) that trap magnetic flux (fluxoids) so that the superconductor sample, at less than 95 K, becomes a permanent magnet. In electrical engineering terms this represents a significant advance in so far as a new powerful source of magnetic field has been made available.

A major difficulty in manufacturing these RE-Ba-Cu-O magnets arises as a result of their complex crystal structure consisting of a $REBa_2Cu_3O_y$ matrix, where y is an integer. Within this matrix there are distributed particles of RE_2BaCuO_y. During processing, the material undergoes a crystalline tetragonal to orthorhombic transformation that introduces deformations in the structure. In addition, oxygen is drawn into the crystal at the melt stage, which subsequently forms microvoids and microcracks, during shrinkage on cooling [15]. These crystal flaws are difficult to prevent, and they become a source of mechanical weaknesses. Thus, the tensile strength of bulk Y-Ba-Cu-O (YBCO), for example, along one of its major planes, is relatively low at between 10 and 30 MPa [16]. As we have already observed, Lorentz forces mean that once a trapped field with its associated shielding current is formed, outward pressure on the crystal proportional to the trapped field is induced. It is not difficult to show [16] that the tensile strength (σ_B) of the YBCO sets the maximum trapped field (B_{max}) that can be tolerated without fracture. The relationship between them is given approximately by

$$\sigma_B = 0.282 B_{max}^2 \quad \text{MPa} \tag{6.26}$$

with B_{max} in Tesla. The implication of this equation is that the maximum trapped magnetic field within an untreated bulk superconducting magnet ranges from 6 to 9.4 T for rare earth compound such as YBCO. Post-fabrication treatment of YBCO involving resin impregnation, and sheathing in carbon fibre to improve mechanical strength, plus drilling and filling with Bi–Pb–Sn–Cd alloy aided by aluminium wire supports to improve thermal conductivity and stability [16] boosts trapped field strengths in a bulk sample of YBCO to 17.24 T at 29 K. As of early 2012, this is about the best that has been achieved. A method for manufacturing superconducting permanent magnets has been granted United States patent [17] as recently as 2006, possibly presaging an increasing availability of these devices for a range of novel applications (see Chapter 7).

Improved magnetic field levels can also be achieved with conventional electromagnets by introducing superconducting solenoids. For high coercive force magnets of conventional construction, of both the ferrite and rare earth varieties, remanent magnetic flux densities in the 1–2 T range can be sustained in closed magnetic circuits. However, at air gaps where high reluctance is experienced, the available flux levels can be considerably lower (see Chapter 4). To compensate for the deleterious effect of the gap, an electromagnet can be employed, as indicated in Chapter 4, and by adjusting the solenoidal current the gap flux densities can be restored to the 1–2 T range for conventional (usually copper) coils, but to 6 T for superconducting solenoids. Unfortunately, the latter exhibits the major drawback of high manufacturing costs.

6.3.4 Recent developments

Since the discovery of type II superconductors, ensuing developments have been largely an exercise in experimental chemistry and material physics. Advances have been sporadic. The main aim has been to find alloys, or compounds, that superconduct at high temperature (relative to absolute zero in the Kelvin scale), and are tolerant to high levels of magnetic flux [1,2,3,13]. The progression is summarised in Figure 6.9 that has been reconstructed from reference 18. Over a period of 60 years from the discovery of superconductivity in mercury, the highest temperature superconductor, which had been found based on metals believed to favour superconduction,

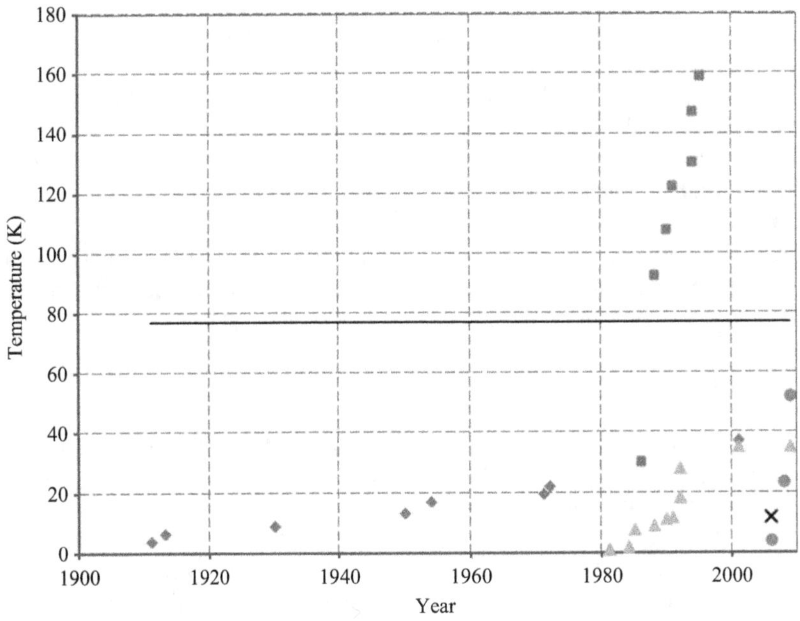

Figure 6.9 Progress in raising critical temperatures of superconducting materials
Diamonds = conventional materials from mercury on left
Triangles = organic materials
Squares = copper oxide alloys from BaLaCuO at low temperature to $HgBa_2Ca_2Cu_3O_8$ at highest temperature
Circles = iron arsenide materials
Cross = fullerene/graphene
Line = boiling point of liquid nitrogen (77 K, −196°C) the dividing line between low and high-temperature superconduction

Table 6.2 Critical temperature (T_c), crystal structure and lattice constants of some high-T_c superconductors

Formula	Notation	T_c(K)	No. of Cu–O planes in unit cell	Crystal structure
$YBa_2Cu_3O_7$	123	92	2	Orthorhombic
$Bi_2Sr_2CuO_6$	Bi-2201	20	1	Tetragonal
$Bi_2Sr_2CaCu_2O_8$	Bi-2212	85	2	Tetragonal
$Bi_2Sr_2Ca_2Cu_3O_6$	Bi-2223	110	3	Tetragonal
$Tl_2Ba_2CuO_6$	Tl-2201	80	1	Tetragonal
$Tl_2Ba_2CaCu_2O_8$	Tl-2212	108	2	Tetragonal
$Tl_2Ba_2Ca_2Cu_3O_{10}$	Tl-2223	125	3	Tetragonal
$TlBa_2Ca_3Cu_4O_{11}$	Tl-1234	122	4	Tetragonal
$HgBa_2CuO_4$	Hg-1201	94	1	Tetragonal
$HgBa_2CaCu_2O_6$	Hg-1212	128	2	Tetragonal
$HgBa_2Ca_2Cu_3O_8$	Hg-1223	134	3	Tetragonal

was a niobium/germanium alloy (Nb3Ge) transitioning at ~20 K. Surprisingly, substantial progress really only came about by a switch to metal oxides, the so-called perovskites (Table 6.2), which would normally be employed as insulators. Fevered experimentation between year 1885 and 1993 pushed T_c for copper oxide compounds from 30 K to 150 K, although this latter figure is for high pressure. In recent years, research into superconductivity in materials that are readily available such as organic compounds and compounds containing iron and arsenic has proceeded apace with transition temperatures in the 30–50 K range being reported.

6.3.5 Fullerenes and graphene

Relatively new arrivals, on the superconducting scene, are the fullerenes [3] and graphene [19]. These carbon atom lattice structures that have evolved from nanotechnology have presented scientists and engineers with tiny 'building-blocks' that are a source of novel materials with surprising and useful properties. The versatility of carbon emanates from the fact that its atoms bond very strongly to each other and to many other elements. As we know they form short molecular chains in gases such as methane and long chains in materials such as plastics and proteins. But they also bond in a manner that produces regular and symmetric geometrical lattices thereby creating very hard materials such as diamond.

Although carbon, in the form of fullerene structures or buckyballs (C_{60}), is created in candle soot, these more exotic structures became seriously important to material scientists once it was discovered that they could be constructed using nanotechnology. The carbon atoms are linked together 'mechanically' into tiny spheres in the case of buckyballs, and into cylinders in the case of nanotubes. A buckyball is about 1 nm in diameter. Graphene, on the other hand, is essentially a two-dimensional molecular sheet being just one atom thick. The crystal takes the form of a hexagonal honeycomb-like lattice. It has become quite well known now that these nanostructures are very light, and exceptionally strong mechanically with tensile stress levels about 100 times that of steel. Electrically they are good conductors – about 10 times higher conductivity than silver – but they are not superconductors without introducing other elements into the lattice.

For the fullerenes, superconductivity has been shown to be possible by lacing the spaces in a three-dimensional buckyball lattice (like footballs in a carrying net) with alkali metals such as potassium (K) or caesium (Cs). Experimentation with potassium, calcium and caesium has progressed to the point where a Cs_3C_{60} compound has displayed superconduction at 37 K (see Figure 6.9). In the case of graphene research chemists have recently boosted superconduction by doping to a high level using potassium or calcium or both [19]. This is essentially achieved by introducing the dopants to both surfaces of a graphene sheet. Of these graphite-intercalation compounds, as they are termed, CaC_6 has recently demonstrated superconductivity up to a critical temperature of $T_c = 11.6$ K, as is shown in Figure 6.9.

6.4 Summary

A recapitulation of the essential characteristics of superconductors that are of relevance to electrical engineers, leads to the following observations:

- In electrical engineering electrons can generally be treated as particles. In superconductors this is not so, and the wave/particle duality associated with quantum mechanics has to be accommodated.
- If the electron is compared to a radar pulse, which is wave-like in its scattering and diffraction characteristics, yet particle-like in exhibiting a discrete position and velocity in space, the wave/particle duality is less difficult to comprehend.
- If a narrow electromagnetic radar-like pulse is directed into a waveguide ring resonator, it is converted into a coherent wave or steady single frequency mode in the ring. In quantum mechanical terms the wideband pulse has become quantized.
- In a single turn inductor, formed from a superconducting wire, similar behaviour occurs where the electron particles cohere into quantized electron waves and these waves pass through the ionic lattice of the bulk material unattenuated – hence exhibiting zero resistance.
- Superconduction is limited by a critical temperature and a critical external magnetic field.
- Superconductors come in two types. Type I are essentially pure elements with almost total magnetic field exclusion (Meissner effect) below the critical temperature and magnetic field values. Type II materials are usually compounds with two critical fields H_{c1} and H_{c2}. Below H_{c1} the Meissner effect is total; above it but below H_{c2} it is partial, with magnetic field fluxoids scattered in a regular pattern through the material.

- Flux pinning is essential to prevent fluxoid drift, and hence resistive loss, due to Lorentz pressure.
- High-temperature superconductivity (above 77 K – boiling point of liquid nitrogen) has been observed only in type II materials.
- Magnetic flux trapping in superconductors, by cooling a melt in the presence of a strong field has been demonstrated, so forming permanent magnets.
- Doped fullerenes and graphene have been shown to superconduct.

Given the recent rate of progress in the development of high-temperature superconducting materials, it is surely not too much outside the bounds of reality to suggest that superconductors operating at normal atmospheric temperatures could be attainable in the not too distant future. In such a case major improvements in the efficiency of electrical devices becomes possible by introducing zero resistance conductors wherever possible – from integrated circuits to the electrical grid.

6.5 References

[1] Tinkham, M. *Introduction to superconductivity*. New York, NY: Dover Publications Inc.; 1996

[2] Lehndorff, B.R. *High-T_c superconductors for magnet and energy technology*. Berlin: Springer-Verlag; 2001

[3] Moon, F.C. *Superconducting levitation*. New York: John Wiley & Sons Inc.; 1994

[4] Vanderlinde, J. *Classical electromagnetic theory*. New York, NY: John Wiley and Sons Inc.; 1993

[5] Jackson, J. *Classical electrodynamics*. New York, NY: John Wiley and Sons Inc.; 1998

[6] Feynman, R.P., Leighton, R.B., & Sands, M. *Lectures on physics II*. Reading, MA: Addison-Wesley Publishing Co., 1964

[7] Kraus, J.D. *Electromagnetics*. New York, NY: McGraw-Hill Book Co.; 1984

[8] Rao, N.N. *Elements of engineering electromagnetics*. New Jersey, NJ: Prentice-Hall Inc.; 1994

[9] Sadiku, M.N.O. *Elements of electromagnetics*. New York, NY: Oxford University Press Inc.; 1995

[10] Yariv, A. *Quantum electronics*. Singapore: John Wiley & Sons Inc.; 1989

[11] Peskin, M.E., & Schroeder, D.V. *An introduction to quantum field theory*. Boulder, CO: Westview Press; 1995

[12] Mead, C.A. *Collective electrodynamics*. Cambridge, MA: MIT Press; 2000

[13] Blundell, S. *Superconductivity*. Oxford: Oxford University Press; 2009

[14] Anderson, P.W., & Kim, Y.B. 'Hard superconductivity: theory of the motion of Abrikosov flux lines'. *Review of Modern Physics*. 1964;**38**:39–45

[15] Tomita, M., & Murakami, M. High temperature superconducting bulk magnets that can trap magnetic fields over 17 T at 29 K. *Nature*. January 2003;**421**: 517–520

[16] Sakai, N., Seo, S.J., Inoue, K., Miyamoto, T., & Murakami, M. *Advances in superconductivity*. Tokyo: Springer-Verlag; 1999, pp. 685–688

[17] Oka, T., Noto, T., & Yokoyama, K. *Superconducting permanent magnet*. US Patent, US 2006/0252650A1, November 9, 2006

[18] Blundell, S. 'Superconductivity – New Scientist Instant Expert 17'. *New Scientist*, November 5, 2011

[19] McChesney, J.L., Bostwick, A., Ohta, T., Seyller, T., Horn, K., González, J., et al. 'Extended van Hove singularity and superconducting instability in doped graphene'. *Physical Review Letters*. 2010;**104**(136803):1–4

Chapter 7

Diamagnetic and eddy current levitation

Engineers . . . are not mere technicians and should not approve or lend their name to any project that does not promise to be beneficent to man and the advancement of civilization.

John Fowler

7.1 Diamagnetism

7.1.1 Conventional elements

In Chapter 3, we discovered that all materials, not just those that are overtly magnetic such as iron, are governed, in classical electromagnetism, by the fundamental relationship

$$\mathbf{B} = \mu_0[\mathbf{H} + \mathbf{M}] \tag{7.1}$$

where the vector field **B** is the internal magnetic flux density in Tesla, generated by an applied magnetic field **H** (A/m) and the magnetic moment per unit volume **M** (A/m) (see Chapter 3). **M** expresses the tendency of the electron spins and orbits within the material to align and enhance (paramagnetic) or counteract and diminish (diamagnetic) the applied field. Again classically we choose

$$\mathbf{M} = \chi \mathbf{H} \tag{7.2}$$

which defines χ, the magnetic susceptibility. The consequence is that the relative permeability of any material is given by

$$\mu_r = \mu_0(1 + \chi) \tag{7.3}$$

Values of χ for some typical diamagnetic materials, commonly used in engineering applications, are provided in Table 7.1.

7.1.2 Compounds

Diamagnetism is actually a general phenomenon, because all electrons, including the electrons of an atom, if viewed as microscopic solenoids (see Chapter 3) must inevitably respond according to the dictates of the Lenz Law to any applied magnetic field. However, for materials that display paramagnetism, and in particular

Table 7.1 Conventional diamagnetic materials – susceptibility χ

Water	-0.91×10^{-5}
Copper	-1.0×10^{-5}
Carbon (graphite)	-1.6×10^{-5}
Lead	-1.8×10^{-5}
Carbon (diamond)	-2.1×10^{-5}
Silver	-2.6×10^{-5}
Mercury	-2.9×10^{-5}
Bismuth	-16.6×10^{-5}
Pyrolitic carbon	-40×10^{-5}
Superconductor	-1.0

ferromagnetism, at the crystalline level, the diamagnetism is swamped by this more powerful phenomenon. In engineering, materials that are classed as diamagnetic are those that are more usually classed as *non-magnetic*. These include many metals such as copper, and particularly the heavy ones with electron dense atoms, such as mercury, gold and bismuth. Most organic compounds, since they are composed mainly of water, such as wood and biological materials, are diamagnetic, as is petroleum, but in this case, it is so because of its carbon content. The truth of the former statement is dramatically illustrated by the now well known photograph, at least to anyone who has delved into magnetic levitation, of a frog floating in air above a powerful magnet (see Figure 7.1). This seemingly weird experiment resulted in an Ig Nobel physics

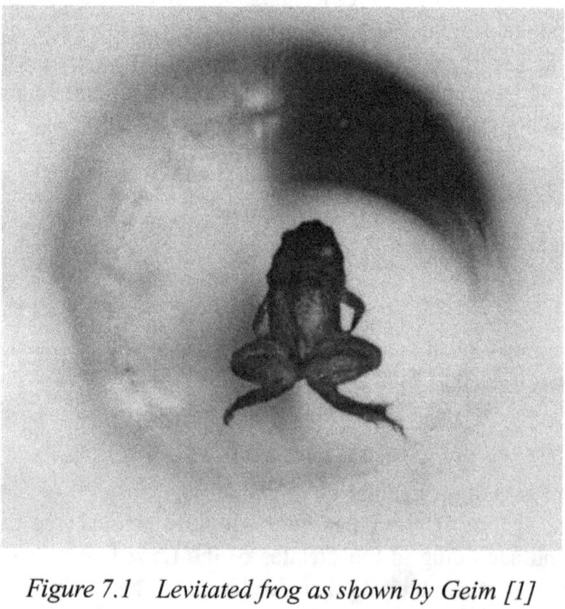

Figure 7.1 Levitated frog as shown by Geim [1]

award for A.K. Geim of the University of Nijmegen, Netherlands, and M. Berry of Bristol University in the UK. The unembellished, matter of fact citation for the award was 'for using magnets to levitate a frog'. Geim also won a genuine Nobel prize, for research on graphene, indicating that his research on magnetism was anything but insubstantial and transitory. He is the only scientist to have been presented with both a Nobel and an Ig Nobel prize.

It is clear from (7.1), and the susceptibility values listed in Table 7.1, that diamagnetic materials have a relative magnetic permeability that is generally just slightly less than unity. For example, copper has $\mu_r = 0.99999$. As the frog experiment demonstrates, diamagnetic materials are repelled by magnetic fields although this effect is hardly observable in everyday life since diamagnetism is so insignificant, as Table 7.1 attests, and the magnitude of a magnetic field required to float a frog is far beyond easily generated levels. The most strongly diamagnetic naturally occurring material is bismuth, $\chi = -1.66 \times 10^{-4}$, although pyrolitic carbon is attributed with a susceptibility of $\chi = -4.00 \times 10^{-4}$ in one plane. Nevertheless, these values are orders of magnitude smaller than the magnetism exhibited by paramagnets and ferromagnets.

While diamagnetic materials resist a changing magnetic field, as required by the Lenz law, by atomic current alignments at the macroscopic level, the law also manifests itself in conductors at the macroscopic level. As we have already noted briefly in Chapter 3, in a metallic plate threaded by a changing magnetic field, current loops are formed in accordance with Faraday's law (see Chapter 2), and these currents are in a direction that tends to generate a negating magnetic field. The natures of these induction currents or eddy currents are further developed in section 7.2 with the aim of providing, later in the Chapter, a gradual introduction to the study of the diamagnetism inherent in superconductors, particularly in relation to levitation roles.

7.2 Eddy currents in metals

7.2.1 Conventional conductors

Diamagnetism manifests itself in all conductors when they are subjected to a magnetic field that is changing with time. Depending on your view point, the 'free' electrons in the conductor are influenced by Faraday induction or the Lorentz force, causing them to circulate around the applied magnetic field lines thus forming eddy current loops (see Section 3.4). The direction of the induced currents is dictated by the Lenz law as we have seen. The eddy currents are mandated to produce an induced magnetic field that opposes the applied field. Repulsion systems employing eddy currents can be classified either as fixed (i.e. no relative movement) alternating field systems or as systems employing motion between a (usually) permanent magnet and a conductor [2,3]. In this section, we will concentrate on the former.

In alternating current (AC) systems, levitation forces are generated by employing electromagnets powered from AC sources (Figure 7.2). Since the eddy currents in the float change direction in sympathy with the field developed by the electromagnet, a steady uni-directional levitation force is formed. The basic principle can be described

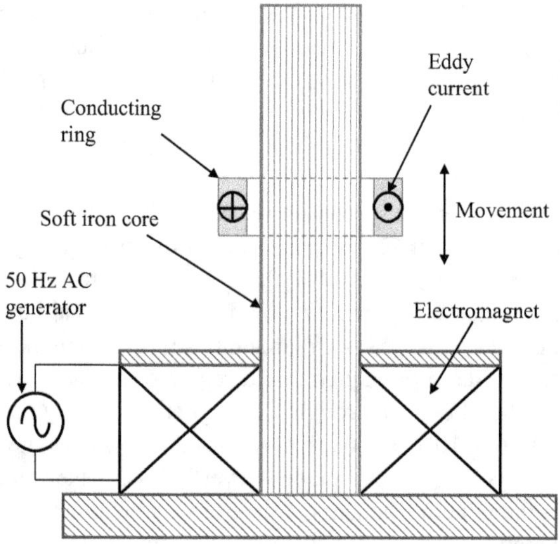

Figure 7.2 Levitation using eddy currents

relatively easily by considering a well-established demonstration designed for the edification of engineering students.

The schematic in Figure 7.2 is reasonably representative of a typical experiment that is often performed in science laboratories to illustrate the principle of eddy current levitation. It comprises an aluminium or copper ring threaded onto a cylindrical soft iron core formed from steel wires, to minimise the induction of disruptive eddy currents within the core itself. The core is embedded at its lower end within a solenoid driven from a 50 Hz AC current source. On switching on the electromagnet the ring rises rapidly up the core to a point where the eddy current levitation force is balanced by the force of gravity. The time-varying field in the core produced by the electromagnet induces eddy currents in the lower surface of the ring, and these circulating currents generate a magnetic field that opposes the original field. This Lenz's law behaviour results in a force of repulsion that lifts the ring. It should be noted that the ring has to remain in contact with the core during the experiment, otherwise it deviates laterally out of the levitating field. In other words, this simple system is totally unstable under the influence of any lateral forces. The ring immediately drops back onto the top surface of the electromagnet when the AC generator is switched off. Any residual DC magnetic flux in the core has no effect on it.

7.2.2 Lateral displacement

The instability alluded to in the levitated ring experiment can be explained by examining the Lorentz force on moving charge in a thin conducting disc located above, close to and parallel to, a simple AC electromagnet as suggested in Figure 7.3.

In Figure 7.3a, the conducting plate is located symmetrically above, and parallel to, the face of a solenoidal electromagnet that is embedded within a laminated iron

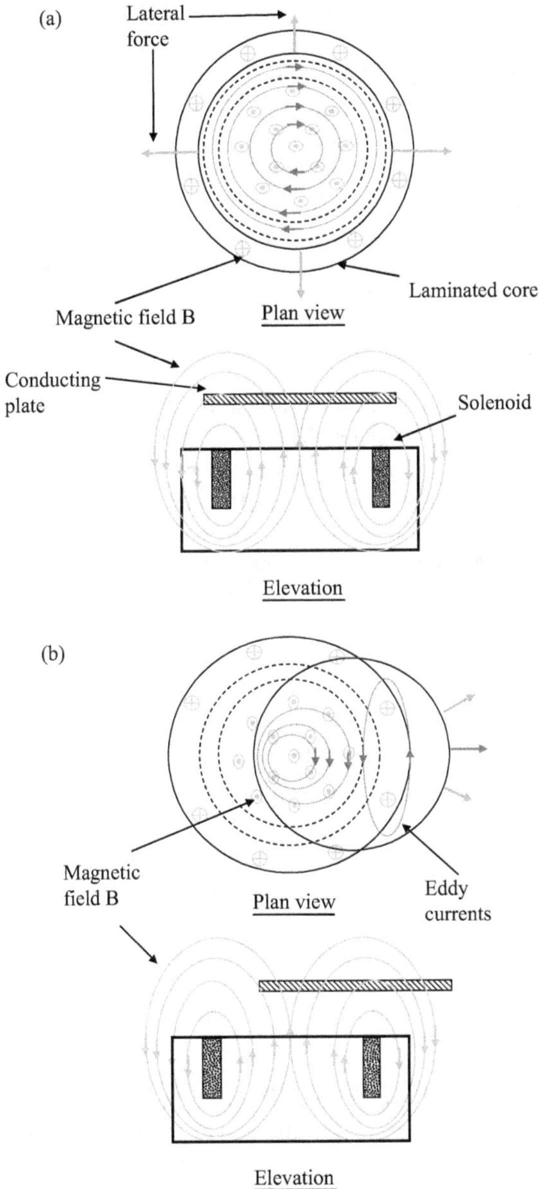

Figure 7.3 Lateral instability associated with eddy current levitation

core. The approximate magnetic field produced by the magnet, at an arbitrary instant in time, is shown by the blue ellipses in the elevation view and by 'arrow heads and fins' in the plan view. These fields penetrate the conducting disc, and since the fields are changing with time, eddy currents (depicted as lightly dotted traces) are induced that encircle the magnetic flux. The currents are in a direction that produces a magnetic field distribution opposing the original field. This can be determined by applying 'clockwise' or 'right-hand' rule to the direction of increase of the magnetic field (assumed to be upwards in Figure 7.3a). The circulating eddy currents can be construed as forming a 'solenoid' located coaxially with the electromagnet (cf. Figure 3.10). Hence, the levitation force can be estimated from the magnetic pressure, much as described in Chapter 3, for the equivalent solenoidal arrangement in Figure 3.10.

Current flow within a conductor is, of course, represented by 'free' electrons in motion, as suggested earlier. If the electron charge is q_e and the tangential velocity at some radius (r) is $\mathbf{v}$, then at that radius, a radial Lorentz force (F), given by (see section 3.4)

$$\mathbf{F} = q_e \mathbf{v} \times \mathbf{B} \quad \text{(Newtons)} \tag{7.4}$$

is formed. However, symmetry dictates that, in the plane of the disc, this force will apply equally in all directions independent of angle, which means that there is no net effect on the disc. In theory, it should float on the applied magnetic field in perpetuity if undisturbed – an implausible scenario in the real world of noise, vibration and thermal air movement.

A disc disturbed from its equilibrium position is suggested in Figure 7.3b. Induction still produces currents in the disc circulating around the changing magnetic field emanating from the electromagnet, but these are now located principally towards the edge of the disc. The Lorentz forces have become unbalanced producing a net force directed to the right in Figure 7.3b. This is clearly underlined by examining the eddy current flow towards the right side of the disc that shows current lines, which are essentially unbalanced by any counter-directed current, resulting in the observed displacement force. The Lorentz force on the moving electrons considered *as a whole* translates to the disc itself because the average rightwards electron shift, within the initially stationary disc, imposes an electrostatic force on the fixed ions forming the metal structure. That is, the 'centre of gravity' of the free electronic charge moves rightwards electrostatically pulling the disc as a whole with it. In practice, once misalignment between the disc and the electromagnet occurs, the disc is rapidly ejected from the magnetic field.

7.2.3 Stabilisation

Disc ejection in a circularly symmetric eddy current levitator, as described above, can in principle be countered by employing two or more concentric coils embedded in the soft iron core, as depicted in Figure 7.4. The inner coil carrying current I_2 provides the levitation fields, while the outer coil (I_1) provides stability in the event of lateral displacement. Stability generally requires $I_1 < I_2$ and I_1 lagging I_2 in phase by approximately 0–45° (see Figure 7.5).

Referring back to Figure 7.3b, the unbalanced eddy currents towards the right edge of the disc are no longer immersed in a downward leakage field (at the instant shown there) resulting in a Lorentz force that reinforces the rightwards movement. Now the stabilising coil supporting a current that is high enough to reverse the leakage fields but not so high that it influences the levitation mechanism establishes a Lorentz

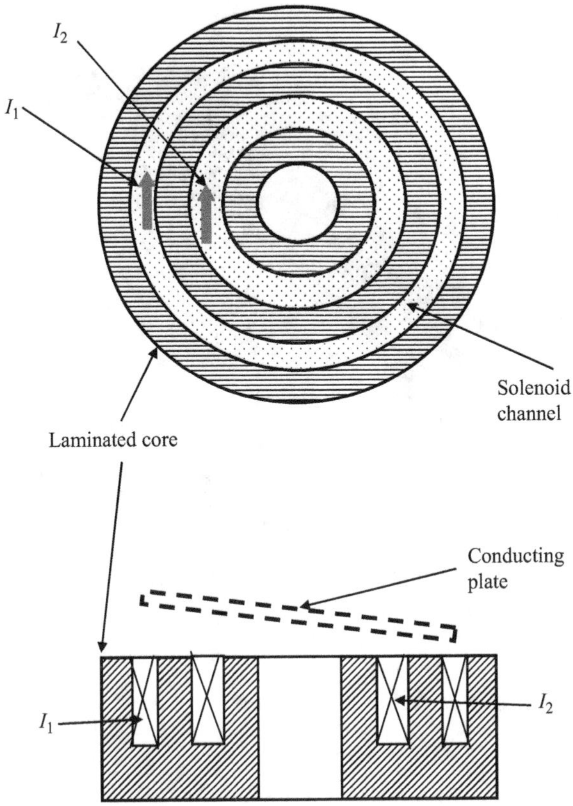

Figure 7.4 Eddy current levitation of conducting disc with stability coil

force, which tends to restore the disc to its equilibrium position. This is true for a range of I_1 current levels, providing I_1 is not too different in phase to I_2 (see Figure 7.5). For this single stabilising coil system, stability can be achieved with phase deviations ranging from a 45° lag to a −45° lead.

Figure 7.5 is based on measurements [3] on a thin conducting disc levitated by a twin coil laminated core with a depth of 14 cm and a diameter of 20 cm. In the recorded experiment, the inner coil comprised 480 turns of copper wire carrying a current of 8 A, while the outer was wound with 440 turns. The figure demonstrates that lateral stability for a level disc is achievable for outer/inner current ratios in the range 0.1–0.6 (black region). On the fringes of this region (light grey shading), stability is still available but only with a slight tilt on the disc. Outside of the shaded regions the disc cannot be levitated stably.

In addition to flat discs, levitation based on eddy currents can also be procured with cylinders and spheres [3], using stabilisation techniques not too dissimilar to what has just been described. The predominant practical application arguably lies in handling molten metal, in what are now rather well-established refining processes [4,5].

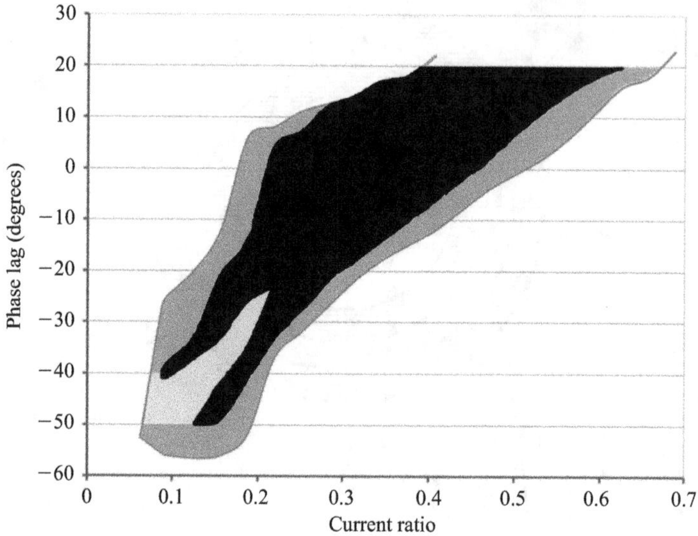

Figure 7.5 Eddy current levitated disc: stability characteristics
Black area = stable
Dark grey area = stable but tilted
Light grey area = disc oscillates
White area = unstable

7.2.4 Levitation through motion

In Chapter 3, it was observed that a DC current carrying solenoid behaves like a permanent magnet with a north and a south pole. Consequently, a pair of coaxial DC solenoids will repel or attract depending on the location of the poles. On the other hand, if one of the coils is merely a passive conducting ring, no interaction occurs between the DC current carrying coil and the passive ring, because a field rate of change is absent. To procure a repulsive (levitating) force, without introducing current variation in the primary coil, it is necessary to introduce relative movement to incur a field change in the vicinity of the ring, so inducing eddy currents. This process lies behind virtually all magnetically levitated transport systems. The technical literature is replete with papers [6–9] that provide theoretical treatments enabling the calculation of levitation forces generated in 'simple' low- and high-speed electrodynamic transportation system. Here we will provide a summary of these developments. For more complex systems, which reflect the reality of practical vehicles, the basic mathematical equations do not change. However, the mathematical complications, engendered by elaborately shaped electromagnets and rails, call for sophisticated 'book keeping' procedures, which can only be handled effectively by resorting to a high-speed computer.

As always the basic equations in magnetic systems are founded in the Maxwell equations. Here we employ the basic integral form of Maxwell's equations, as outlined

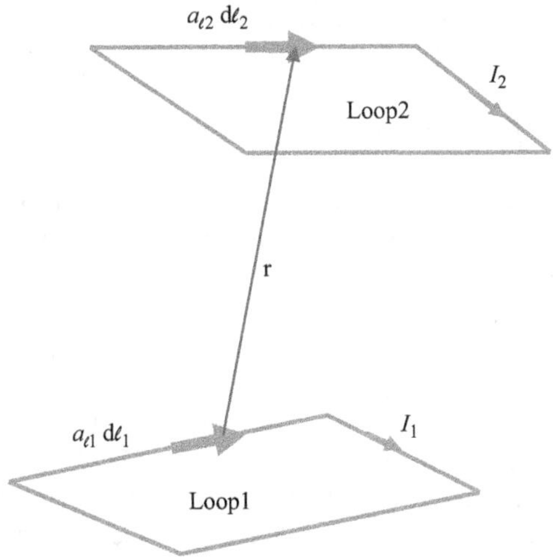

Figure 7.6 Lorentz force between current loops

in Chapter 2, to show that the Ampere Law when applied to a current element $I\,\mathrm{d}\ell$, which forms part of a current loop in an electrodynamic levitation scheme (Figure 7.6), yields the following relationship for the magnitude of the magnetic field at a point in space due to the element (see (2.31)):

$$\therefore\ \mathrm{d}B = \frac{\mu_0 I\,\mathrm{d}\ell}{4\pi r^2} \tag{7.5}$$

However, the relationship between current and magnetic field is vectorial and this needs to be accommodated into the equation. This can be done by expressing the direction of the current element through the unit vector $\mathbf{a}_\ell$ and the direction of the radial vector ($\mathbf{r}$) by introducing the unit vector $\mathbf{a_r}$. Equation (7.5) now becomes, for current element $I_1\mathrm{d}\ell_1$ (referring to Figure 7.6)

$$\mathrm{d}\mathbf{B} = \frac{\mu_0 I_1\,\mathrm{d}\ell_1\mathbf{a_r}\times\mathbf{a}_{\ell 1}}{4\pi r^2} \tag{7.6}$$

At $\mathrm{d}\ell_2$ on loop 2, the presence of field $\mathrm{d}\mathbf{B}$ means that a Lorentz force exists given by

$$\mathrm{d}\mathbf{F} = I_2\mathrm{d}\ell_2\mathbf{a}_{\ell 2}\times\mathbf{a}_r \tag{7.7}$$

Combining (7.6) and (7.7) results in the differential relationship:

$$\mathrm{d}\mathbf{F} = \frac{\mu_0 I_1 I_2\,\mathrm{d}\ell_1\mathrm{d}\ell_2}{r^2}\mathbf{a}_{\ell 1}\times\mathbf{a}_{\ell 2}\times\mathbf{a_r} \tag{7.8}$$

The total Lorentz force between loop 1 and loop 2 in Figure 7.6 is obtained, in essence, by summing up the effects of all elements $I_1\mathrm{d}\ell_1$ that make up loop 1, at all $I_2 d\ell_2$ that completes loop 2. If the elements are small enough, the result is an integration, leading to [10]

$$\mathbf{F} = \frac{\mu_0 I_1 I_2}{4\pi} \oint_{\ell 1} \oint_{\ell 2} \frac{d\ell_1 d\ell_2}{r^2} \mathbf{a}_{\ell 1} \times \mathbf{a}_{\ell 2} \times \mathbf{a}_{\mathbf{r}} \tag{7.9}$$

This is a very general equation, being independent of the relative positioning and orientation of the loops. It permits the computation of both vertical and lateral forces on the coil of interest.

The typical electrodynamic levitation problem can be depicted as comprising a vehicle-mounted solenoid at height h above an electrically conducting track (Figure 7.7), which for simplicity is represented by a conducting sheet. If the current in the vehicle-mounted coil is unchanging with time (DC), no interaction occurs between the coil and the conducting sheet as required by Faraday's law. Only when eddy currents are induced in the sheet are interaction forces developed. These are calculable using (7.6). For a DC electromagnet, the presence of eddy currents is possible only if relative motion between the coil and the sheet exists. If the eddy current layer (track) in Figure 7.7 is perfectly conducting, image theory can be used to compute the 'ideal' lift force between the current plane and the moving coil. This implies introducing an image coil with anti-phase currents at a distance below the plane equal to the distance of the 'real' coil above it. This coil produces eddy currents in the conducting layer that exactly negate the original currents so that the layer can be removed without altering the fields and forces acting on the original solenoid. On applying (7.6) to this

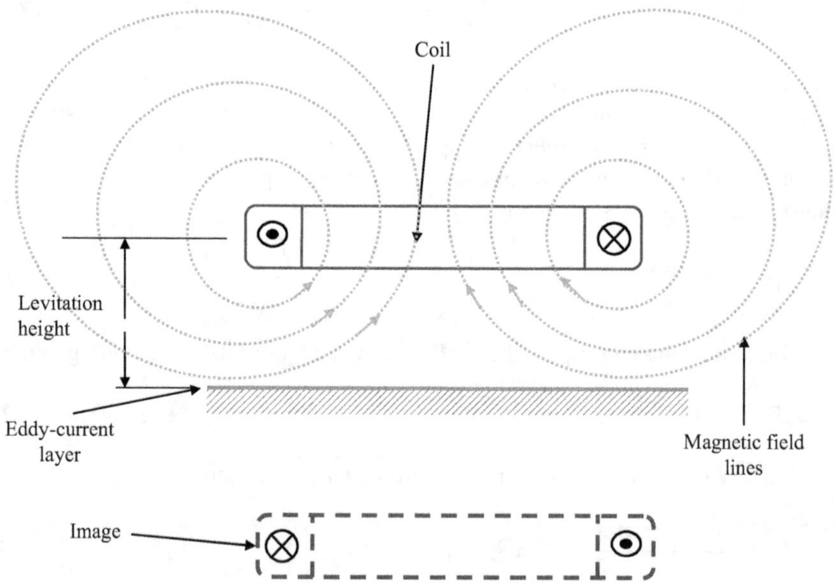

Figure 7.7 Eddy current levitation of a moving coil showing image method of force computation

Table 7.2 Ideal lift between a rectangular coil and a conducting layer (multiply tabulated values by $I_1I_2(a+b)/h \times 2 \times 10^{-7}$ to obtain lift force in Newtons when current is in Amperes)

2h/a	**b/a = 0.5**	**b/a = 1.0**	**b/a = 2.0**	**b/a = 10**
0	1.0	1.0	1.0	1.0
0.1	0.8601	0.9009	0.9316	0.9742
0.2	0.7143	0.8041	0.8601	0.9430
0.3	0.5771	0.7108	0.7870	0.8822
0.4	0.4591	0.6228	0.7143	0.8224
0.5	0.3633	0.5417	0.6438	0.7583

coil and image problem, with a separation of $2h$ between them, yields the ideal lift results shown in Table 7.2 [10]. These apply to a moving rectangular coil length a, and width b, at height h above a conducting sheet. Note that the b dimension lies in the direction of motion. Clearly, a long narrow loop ($b/a = 10$) with its major dimension aligned with the direction of motion is most effective in generating lift at moderate speeds. Fortuitously, this accords with the geometry of rail transport systems.

The 'ideal' lift forces listed in Table 7.2 are limited to scenarios where the eddy currents in the conducting plane – and hence the coil image – are presumed to be located directly below the moving electromagnet. In practice, this is a simplification. This has been demonstrated [8] by solving the differential equation for the magnetic field $\mathbf{B}_e$ associated with the moving eddy current pattern in the conducting sheet (conductivity σ, permeability μ_0). The equation is

$$\frac{\partial \mathbf{B}_e}{\partial t} = v_{\mathrm{TEM}} \frac{\partial \mathbf{B}_e}{\partial \mathbf{z}} \tag{7.10}$$

where v_{TEM} is the velocity of a plane (TEM) electromagnetic wave at frequency ω within the metal. In levitation discussions, it is termed the characteristic velocity. Mathematically

$$v_{\mathrm{TEM}} = \frac{\omega}{\beta_{\mathrm{TEM}}} = \frac{2}{\omega \mu_0 \delta} \tag{7.11}$$

The skin depth δ is given by the relation

$$\delta = \sqrt{\frac{2}{\omega \mu_0 \sigma}} \tag{7.12}$$

The solution in reference 8 indicates that the eddy currents decay in the wake of the moving source, and can be represented by 'a wake of images'. The analytical results are summarised in Figure 7.8 where the lift force (F_L), normalised to the ideal lift force presented in Table 7.2, is displayed as a function of velocity (v) normalised to the characteristic value for the conducting sheet. It shows that lift is negligible, as expected, at low velocities, building towards the ideal value for coil velocities exceeding the characteristic value by a factor of at least 20. At these speeds, the eddy currents

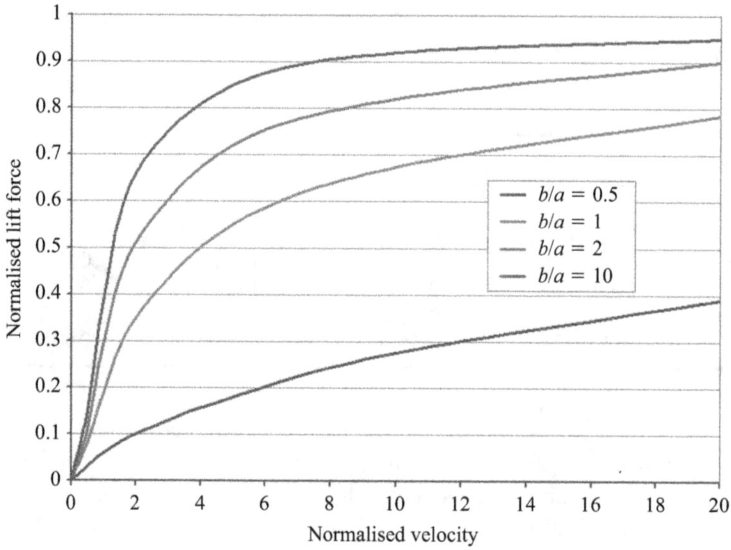

Figure 7.8 Normalised lift force for a moving coil above a conducting sheet (track), as a function of normalised velocity

are well formed. Computations on this relatively simple, but instructive model, also predicts the presence of an electromagnetic (EM) drag force (F_D), particularly at low speeds, with the relationship between drag and lift given by, and generally confirmed by measurements on models:

$$\frac{F_D}{F_L} = \frac{v_{TEM}}{\mathbf{v}} \tag{7.13}$$

It should be noted that the EM drag is much smaller than aerodynamic drag even in a vehicle designed for high speeds. Typically, in a bullet train travelling at 500 km/h, the EM drag could be 1.5% of the lift force, getting larger at lower speeds as predicted by (7.13). Even so, it is still deemed sufficiently troublesome for the emergence of schemes aimed at substantially reducing this drag. Termed 'null-flux' methods, these have been examined in detail in reference 10. The technique essentially involves track arrangements incorporating conductor loops interconnected in a manner that forms anti-phase fields in adjacent loops, hence producing the desired null flux.

Examples of electromagnetically levitated transport systems are plentiful in the academic literature. The most practical and relevant developments are fully addressed and summarised in texts such as those listed in references 2, 3, 10, 11. Here we simply note that the drive to secure low loss, high speed, electrodynamic rail systems is based on computations and modelling, much as described above, and these efforts indicate that coach-mounted electromagnets, delivering not unreasonable flux densities in the 1 T range, can create magnetic pressure levels between train and rail that are more than enough to float, at ~15 cm above an aluminium track, a 100 passenger vehicle weighing around 30 tons. The basic concept is illustrated in Figure 7.9.

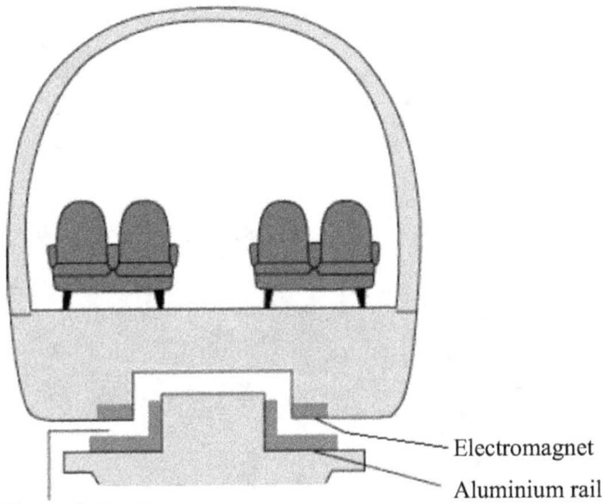

Figure 7.9 Schematic illustration of an electrodynamically levitated rail transport system

7.2.5 Levitation based on linear induction

Thus far, in this chapter it has been observed that levitation associated with eddy currents can be achieved by employing an AC electromagnet in the close vicinity of a conducting rail or by means of a DC electromagnet located close to and in relative motion to a fixed conducting rail. A third method lies in the creation of a spatially moving magnetic field, or more accurately a travelling electromagnetic wave, formed by the rail system that comprises sequentially switched or phased electromagnets. The technique has its genesis in induction motors where a rotor cage constructed from conducting rods is driven by Lorentz forces to spin in synchronism with a rotating magnetic field produced by sequentially switched coils arranged around the fixed armature. The linearised version that has been adopted for use in levitated transport systems has the 'flattened' armature forming the railway system while the planar rotor is housed in the moving vehicle. A major difference between the rail system and the motor is that, in addition to propulsion, the Lorentz forces between the 'armature' and the 'rotor' are employed to provide levitation. The concept, often referred to as the 'magnetic river' [12,13], is simplistically illustrated in Figure 7.10. It comprises 'rails' formed from U-shaped steel pressings. These are stacked to create U-shaped cores and wound to realise the rail elements as shown in the end view in Figure 7.10. With the polarity indicated, lateral stability of a conducting float is assured [12]. Propulsion along the rail system is procured by distributing the magnets periodically along a linear guideway, as suggested in the plan view of Figure 7.10, and introducing current switching or phasing to create a travelling magnetic field pattern (at velocity v_f). Velocity v_p is the terminal float speed that will generally be much slower than v_f.

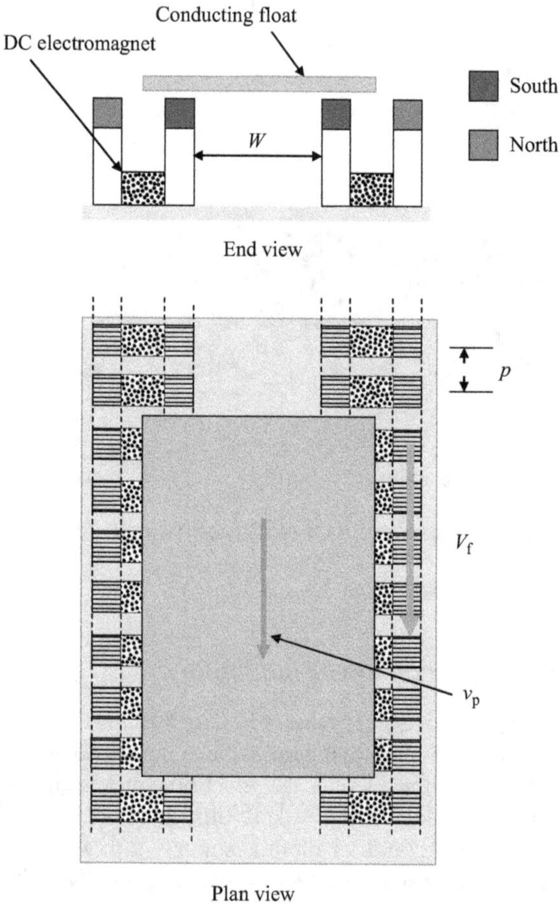

Figure 7.10 Transverse flux levitated rail system with periodic magnet track

Actually, the lift and drive mechanism is not unlike the inductive lift and stabilisation force discussed in connection with Figure 7.4, except that in the linearised version the changing magnetic flux required to induce circulating currents in the float, and hence a lift force, is procured by sequential switching or phase progression of the current excitation along the track. Numbers of magnetic elements are usually treated as a group, this number depending on the axial reach of the float. Unfortunately, the magnetic circuit of each U-shaped element (see Figure 7.11) is extremely inefficient for a non-magnetic conductive float, because the air gap is impractically large and exhibits very high reluctance. This problem is exacerbated by what is known as unbalanced magnetic pull (UMP), the source of which is depicted in Figure 7.11. While the magnetic pressure that permits lift is as described in Chapter 3, namely

$$P_{\mathrm{L}} = \frac{B_n^2}{2\mu_0} \quad (\mathrm{N/m^2}) \tag{7.14}$$

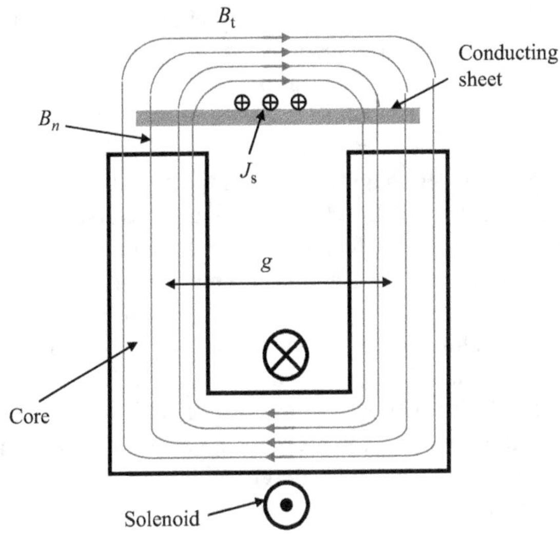

Figure 7.11 Normal forces on flat conducting sheet in 'transverse' field pattern

an opposing force is formed by the tangential component of field at the float, in a transverse flux machine. This 'attractive' pressure arises because $H_t(=B_t/\mu_0)$ at the conducting surface requires an orthogonal surface current J_s $(=H_t)$ to reside there (see Chapter 2). Hence the application of Lorentz law for a current skin on a highly conducting surface (cf. (6.21)), namely

$$\mathbf{F} = \mathbf{J}_s \times \mathbf{B} \tag{7.15}$$

leads to the attractive pressure

$$P_A = \frac{\mu_0 J_S^2}{2} \quad (\mathrm{N/m^2}) \tag{7.16}$$

Hence the total pressure on a float in a levitated linear motor system is

$$P_T = \frac{B_n^2}{2\mu_0} - \frac{\mu_0 J_S^2}{2} \quad (\mathrm{N/m^2}) \tag{7.17}$$

In rotary machines the second term in this equation can usually be deemed to be negligible. On the other hand, in linear propulsion systems where large air gaps (g) are difficult to avoid and where effective drive conditions require many magnetic poles with narrow teeth, the pitch p is restricted. The upshot is that [12]

$$\frac{J_S}{B_n} \approx \frac{g}{p} \tag{7.18}$$

This ratio is no longer small and the second term in (7.17) can become dominant. Consequently, input power to propulsion efficiencies can be very poor. Efficiencies of less than 10% are indicated [12]. In addition, the cost of implementing and maintaining the energised track has been deemed to be prohibitive.

Not surprisingly perhaps, progress on the 'magnetic river' concept was made feasible by inverting the system as shown in Figure 7.12, so that the energised periodic magnet structure becomes an integral part of the moving vehicle (single-sided linear induction motor – SLIM), while at the same time the track becomes much easier to implement as a smooth flat conducting rail embedded within a steel support structure. The presence of steel hugely improves the operational effectiveness of the magnetic circuits. An illustrative representation of a transverse magnetic field levitation system, showing the essential features of this technology, has been provided in Figure 7.12. The obvious cost and implementation gain of transferring the periodic magnetic structure to the moving vehicle is offset by the limited length available. The abruptly terminated travelling wave structure, formed by the phased electromagnets on the vehicle, leads to end effects, and a disruptive standing wave, which severely undermines vehicle stability [13]. As a consequence, a plethora of structural geometries, or enigmatic variations on a theme, have appeared with particular advantages and disadvantages, all of which can be described as SLIM rail systems. These are well described in the literature [2,12–15].

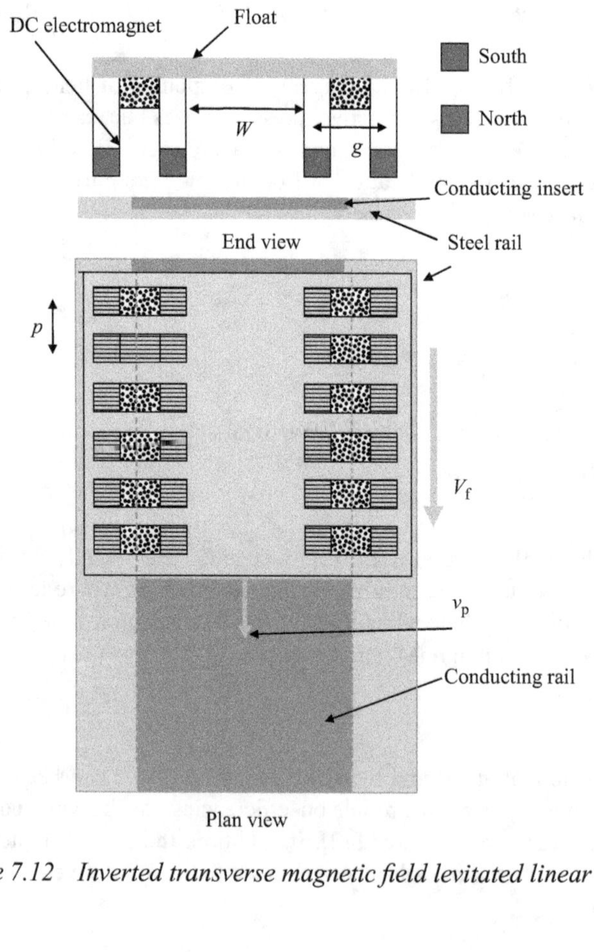

Figure 7.12 Inverted transverse magnetic field levitated linear motor

Much of the passive maglev guideway design and development referred to above was undertaken in the 1970s and early 1980s. The use of a single-sided linear induction motor (SLIM) with a track constructed of iron-backed aluminium held early promise. But for high-speed operation the SLIM was shown to suffer, as we have seen, from a low power factor and vehicles had to be designed to be very long in order to counteract the end-effects [16,17]. In some proposals, a secondary suspension system must be used and this increases costs and creates drag losses [18,19]. When the SLIM track is constructed only of aluminium then the magnets suffer from large gaps of high reluctance, and the induced guideway currents can produce an undesirable attractive force, in addition to lift. Thus, the 'electromagnetic river' concept has an abysmally low power factor [3,13,20,21], and the lift-to-weight ratio is low. Needless to say, it has never been fully implemented beyond modelling level. Interestingly, an electrodynamic wheel, which eliminates end effects, is described in detail by Bird [22,23]. It has been mooted recently as a possible solution to the lingering difficulties of incorporating a linear motor within the vehicle of a rail-based levitation system.

In the context of the search for efficient and cost-effective maglev systems, much of the above historical summary of the development of eddy current solutions seems negative. However, the 'landscape' has been completely reshaped by the advent of high-temperature superconductors. Inevitably, the advent of these revolutionary new materials has 'sparked' major reassessments of the technology. The salient features of this effort will be summarised in section 7.3.

7.3 Diamagnetic levitation with superconductors

7.3.1 Electrodynamic levitation

Superconductors, as indicated in Chapter 6, essentially behave as perfect diamagnets ($\chi = -1$, $\mu_r = 0$), since they expel all magnetic fields from their interior (except in a thin surface layer) due to the Meissner effect. However, this effect is not due to conventional eddy currents as occurs in ordinary metallic conductors, but to cumulative electron wave action when the material is sufficiently cold (see Chapter 6). This perfect diamagnetism of superconductors hugely enhances the repulsion force between a magnet and conductor, and this has reawakened interest in electrodynamic levitation systems such as those depicted in Figure 7.13. There are evidently three ways in which superconduction can be incorporated into an electrodynamic levitation scheme. These are, first, and fairly easily by employing electromagnets wound from superconducting wire (Figure 7.13a and c), second and more challengingly by forming the reaction surface, plate or rail, from a high-temperature superconductor (HTS) (Figure 7.13b and d) or third by employing superconducting rare earth permanent magnets as described in Chapter 6. This option is illustrated in Figure 7.14.

It is perhaps instructive to point out that the magnetic circuits depicted in Figures 7.13 and 7.14, incorporating as they do superconducting elements, differ very little theoretically to equivalent arrangements encountered in systems performing

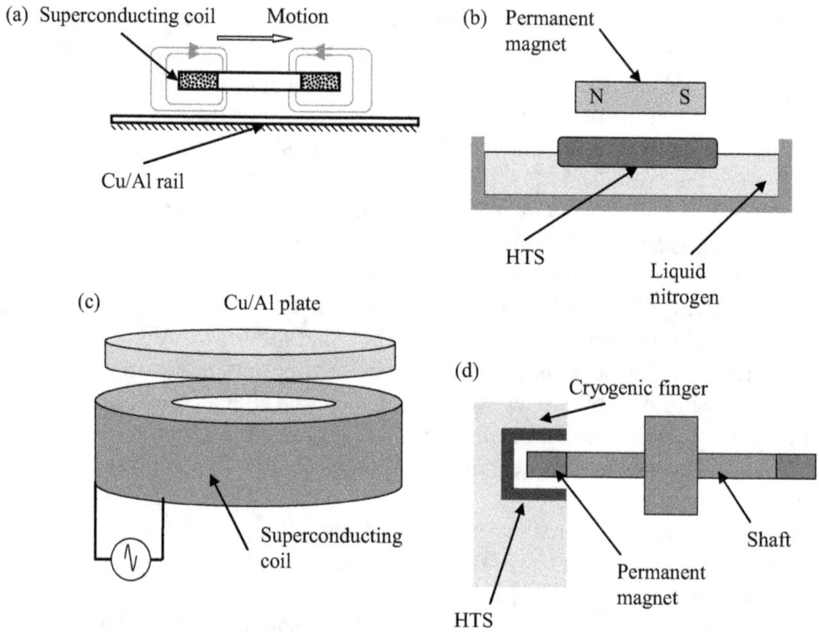

Figure 7.13 Levitated bearing concepts incorporating superconducting (type I) materials

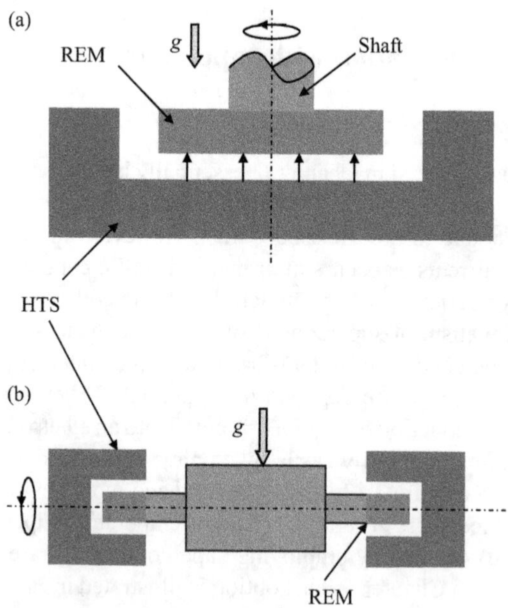

Figure 7.14 Thrust bearing (a) and journal bearing (b) incorporating rare earth superconducting (type II) magnets

similar tasks but using regular components. Consequently, the mathematical equations and the solution techniques outlined in Chapters 3 and 4 can be applied, with care, to levitating devices or systems based around superconductors. This is also largely true of the dynamics of levitated vehicles employing superconducting elements that has already been addressed. It should be noted, however, that the diamagnetic nature of superconductors presents stability advantages because the limitations usually imposed by Earnshaw's theorem are circumvented [24].

For example, the force (F) on the type I, HTS base in Figure 7.15, due to the rare earth cylindrical magnet (length L, diameter D) suspended above by the repulsive force of exclusion, can be estimated by employing (3.15), integrated over the end surface. If the axial magnetic flux density at radius r is B_r, then

$$F = \int_0^{\infty}\int_0^{2\pi} \frac{B_r^2}{2\mu_0} r \, d\varphi dr = 2\pi \int_0^{\infty} \frac{B_r^2}{2\mu_0} r \, dr \tag{7.19}$$

Maximum force obviously occurs at minimum separation, and if we quite reasonably neglect demagnetisation in the rare earth magnet, it is not difficult to show that the pressure in the gap is given by

$$P = \frac{F}{\pi D^2} \approx 0.75 \left(\frac{B_{av}^2}{2\mu_0}\right)\left(\frac{D}{L}\right)^2 \quad (\text{N/m}^2) \tag{7.20}$$

where B_{av} is the average flux density in the 'zero height' gap. The equation clearly indicates that the nature of the type I HTS base is not relevant to the pressure calculation that is predominantly dependent on the magnetic stress on the face of the magnet. This implies that as long as the base is superconducting, the levitation mechanism is largely independent of its material composition. For a rare earth magnet B_{av} could be as much as 0.5 T and hence, depending on the magnet geometry, very practical bearing pressures (P) of the order of 10 N/m^2 are not unrealistic. However, critical temperatures for type I superconductors tend to be impractically low, which mitigates against their use. Furthermore, if operated close to critical, the superconduction state can be quenched by high magnetic fields, as quantified by (6.20). This

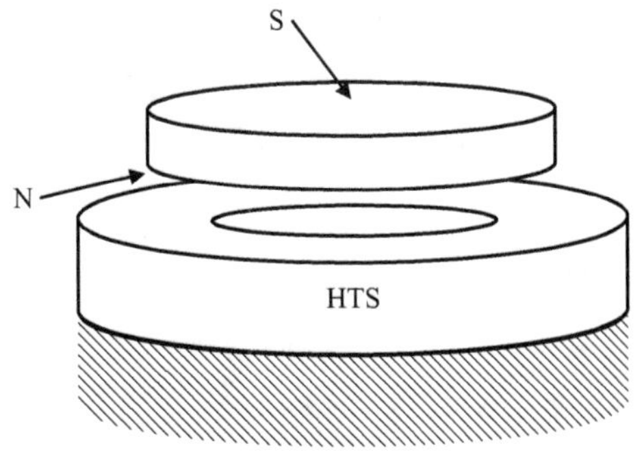

Figure 7.15 Rare earth magnet suspended above a flat superconducting base

represents a distinct drawback in levitation applications. Hence type II materials are more commonly employed in this area.

7.3.2 Type II superconductors

Type II materials exhibit the advantage of high and more manageable critical temperatures but still present engineering problems relating to magnetization. If the base in Figure 7.15 is a type II ceramic superconductor, such as $YBa_2Cu_3O_7$, issues of hysteresis arise [10], which are related to flux pinning (see Chapter 6). In high-temperature superconducting ceramics it has been noted that the induced currents, or the effective magnetization, which they generate, functions differently for an increasing magnetic field as compared to one in decline [25]. In other words, the levitation force behaviour is dependent on the previous history of field applications to the HTS. Consequently, for an arrangement such as that depicted in Figure 7.15, where the applied field varies with the separation distance between the magnet and the superconductor, it follows that the repulsion force tends to display hysteresis as shown in Figure 7.16. For a magnetic float under the influence of gravity, the height at which it hovers above the superconducting base will be conditioned by the past history of movement.

Actually, hysteresis in relation to levitation using type II superconductors is somewhat more complicated than suggested above. Experiments performed on end bearings on a cylindrical shaft [26] aimed at correlating stiffness to bearing configuration have demonstrated the need for care in implementing this technology. The end bearing in one manifestation comprised a shaft-mounted rare earth magnet above a type II superconducting stator (see Figure 7.15). In this case, incrementing

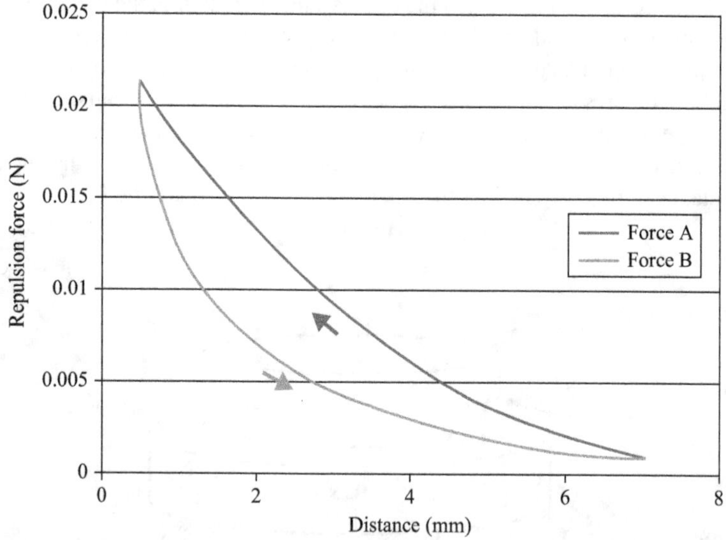

Figure 7.16 Repulsion force versus distance between a rare earth test magnet and a free-sintered YBCO type II superconductor (see Figure 7.15 for geometry)

the gap from zero to 1 mm resulted in the red curve of repulsion force versus displacement as shown in Figure 7.17. A distinct hysteresis loop is in evidence. After the second traverse of the cycle of displacement, the loop becomes repeatable. When the superconductor was replaced by another rare earth magnet in 'N–S–N–S' alignment, the linear negatively sloping curve results with the force being attractive rather than repulsive. Linearity occurs because both high coercive force magnets are fully saturated for all gap sizes. Interestingly, when the stator comprises the superconductor with the rare earth magnet below it, an intermediate trace is generated indicating that the original repulsion force is significantly diminished by the presence of the magnet below it (lightly shaded loop in Figure 7.17). It should be pointed out that this result is not just a simple superposition of the Mag–HTS and the Mag–Mag cases [26].

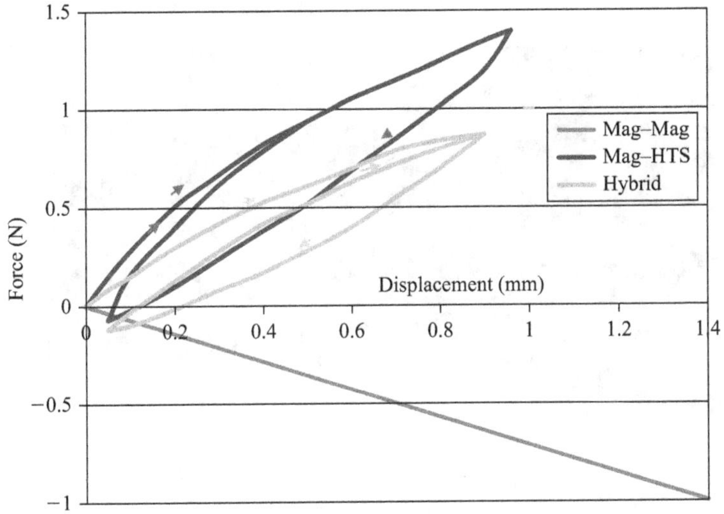

Figure 7.17 Force versus displacement in a composite end bearing comprising (a) opposed permanent magnets on stator and rotor (negatively sloping line); (b) high-temperature superconductor in place of stator magnet (darkly shaded loop); (c) high-temperature superconductor sandwiched between stator and rotor magnets (lightly shaded loop) [26].

The behaviour, represented by Figures 7.16 and 7.17, has implications for bearing 'stiffness' (see Chapter 4) and awareness of it is obviously critical to the accurate design of superconducting bearings.

7.3.3 Superconduction in transport systems

The impact of superconductors on transport systems has advanced significantly by the discovery of high-temperature superconducting materials, and certainly much more than with generally quite compact, low volume, bearings where cooling is less of an engineering issue. The realistic introduction of superconductors to train propulsion and levitation systems was delayed until the late 1980s when materials with T_c values below 95 K were revealed. This meant that the readily available liquid nitrogen (liquid at 77 K) could be employed as a cryogenic agent. By 1991, maglevs employing high-temperature superconductors in both wire and bulk form were on the drawing board, particularly in Japan, although major technical issues remained to be resolved in relation to cryogenics, thermal stability in magnetic systems, superconductor fabrication and magnetic shielding [10].

By 2011 it was possible to assert that a maglev train incorporating superconducting magnets has reached an advanced stage of development in Japan. Powerful onboard superconducting magnets on the train (Figure 7.18) provide frictionless support above the electromagnet-lined tracks. The prototype train shown in the photo has reached a speed of 411 km/h (256 miles/h).

Figure 7.18 Japanese maglev prototype

Running alongside the track are walls (see Figure 7.18 – light grey panels on either side of the vehicle), with a continuous series of vertical coils of wire, mounted inside. The wire in these coils is not superconducting. Cryogenic efficiency dictates that the superconducting magnets must be housed on the train. As the vehicle passes each wall coil, the field of the superconducting magnet induces a current in the adjacent coil, making it a temporary electromagnet generating a field that opposes the vehicle magnets. Thus, the interaction of the superconducting electromagnets on the train with these external guide rail coils produce forces that keep it centered above the track as suggested in Figure 7.19 (right side vehicle representation). At the same time, permanent guide rail magnets, mounted as indicated in the left-hand drawing of Figure 7.19, employ the upwardly directed repulsive and attractive forces between the guide magnets and the super-cooled on-board electromagnets to levitate

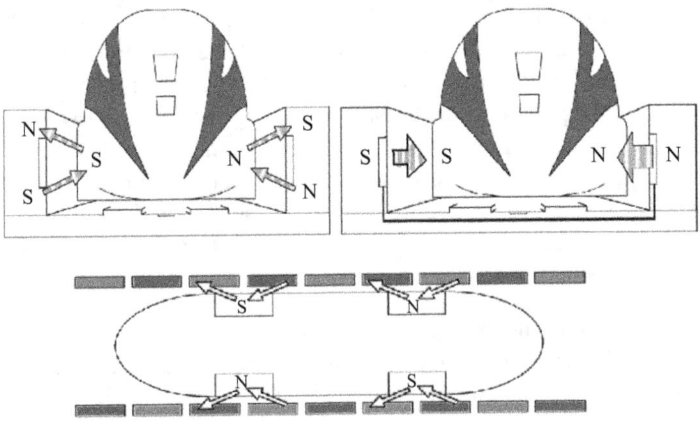

Figure 7.19 Deployment of superconducting coils in Japanese maglev

the train. In addition, in this Japanese system, a phased electric current is made to progress down the linear array of external coils lining the guideway thus providing forward propulsion for the train in much the same way as described in connection with Figure 7.10. This is represented schematically on the plan view of the guide system (see Figure 7.19, lower drawing).

Self-evidently this is not the only development that combines maglev, superconduction and cryogenics, as reference 10 makes clear. The technology continues to evolve spurred on by the recently added goal of achieving all electric high speed transcontinental travel, in a coming world without fossil fuels and with, as a result, very limited air travel.

7.4 Summary

Within the boundaries imposed by our exploration of the technology of levitation and suspension, this chapter addresses diamagnetic phenomena in three categories. These are:

1. *Non-metallic materials*: In this case, diamagnetism is generally a very weak property that is usually swamped by paramagnetic effects. Biological materials are diamagnetic, very graphically demonstrated by a now famous experiment in which a frog is floated in a powerful magnetic field. The physiological repercussions for the frog are not recorded.
2. *Metals*: In a non-magnetic environment 'good' conductors are either weakly diamagnetic or weakly paramagnetic. However, in a changing magnetic field, currents are induced in a conductor. In accordance with Lenz's law, these currents adopt circulation directions that generate secondary fields opposing the original field. In effect, the conductor behaves in a strongly diamagnetic manner. This effect is used widely in levitation systems from bearings to trains.

3. *Superconductors*: These represent the optimum in diamagnetism by exhibiting a magnetic susceptibility of –1. The strength of the levitation forces available with type I superconductors has provided a new impetus to developments in levitation, although it comes with the disadvantage of the need to provide super-cooling. High temperature, type II superconducting materials, within which strong magnetic fields can be trapped, have become a source of very powerful permanent magnets at liquid nitrogen temperatures. This relatively recent development has also given a considerable boost to the search for successful, cost effective levitation systems.

If diamagnetic levitation technology, empowered by superconduction, is to provide a major source of efficiency enhancement in electrical machines, in a renewables' future of restricted levels of power, it is clear that the 'high temperature' at which superconduction can be sustained will need to improve and move closer to normal atmospheric temperatures. The rapid advances in the science of superconduction in recent years, suggests that this is not an unrealistic possibility, and these materials may exist by the time fossil fuel combustion is deemed dangerous for life on Earth.

7.5 References

[1] Simon, M.D., & Geim, A.K. 'Diamagnetic levitation: flying frogs and floating magnets'. *Journal of Applied Physics.* 2000;**87**(9):6200
[2] Rhodes, R.G., & Mulhall, B.E. *Magnetic levitation for rail transport.* Oxford, UK: Clarendon Press; 1981
[3] Jayawant, B.V. *Electromagnetic levitation and suspension techniques*. London: Edward Arnold Publishers Ltd.; 1981
[4] Weisberg, L.R. 'Levitation of molten gallium, indium, gold and tin'. *Review of Scientific Instruments.* 1959;**30**:315
[5] Orkress, E.C., Wroughton, D.M., Comentez, G., Brace, P.H., & Kelly, J.C.R. 'Electromagnetic levitation of solids and molten metals'. *Journal of Physics.* 1952;**23**(12):1413, 1952
[6] Ooi, B.-T. 'A dynamic circuit theory of the repulsive magnetic levitation system' *IEEE Transactions on Magnetics.* 1975;**MAG-11**(5):1495
[7] Miericke, J., & Urankar, L. 'Theory of electrodynamic levitation with a continuous sheet track'. *Applied Physics.* 1974;**3**:67
[8] Reitz, J.R. 'Forces on moving magnets due to eddy currents'. *Journal of Applied Physics.* 1970;**41**(5):2067
[9] Reitz, J.R., & Davis, L.D. 'Force on a rectangular coil moving above a conducting slab'. *Journal of Applied Physics.* 1972;**43**(4):1547
[10] Moon, F.C. *Superconducting levitation.* New York, NY: John Wiley & Sons Inc.; 1994
[11] Sinha, P.K. *Electromagnetic suspension: dynamics and control.* London: Peter Peregrinus Ltd.; 1987

[12] Eastham, J.F., & Laithwaite, E.R. 'Linear induction motors as electromagnetic rivers'. *Proceedings of IEE*. 1974;**121**(10):1099–1108

[13] Laithwaite, E.R., Eastham, J.F., Bolton, H.R., & Fellows, T.G. 'Linear motors with transverse flux'. *Proceedings of IEE*. 1971;**118**(12):1761–1767

[14] Laithwaite, E.R. (Ed.). *Transport without wheels*. Boulder, CO: Westview Press; 1977

[15] Laithwaite, E.R. *History of linear electric motors*. London: Macmillan Book Co.; 1987

[16] Eastham, J.F., Balchin, M., Coles, P., & Rodger, D. 'Comparison of short primary linear machines for high speed maglev vehicles'. *IEEE Transactions on Magnetics*. 1987;**23**(5):2338–2343

[17] Higuchi, T., Nonaka, S., & Ando, M. 'On the design of high efficiency linear induction motors for linear metro'. *Electrical Engineering in Japan*. 1987;**137**(2):36–43

[18] Borcherts, R.H., & Davis, L.C. 'The superconducting paddlewheel as an integrated propulsion levitation machine for high speed ground transportation'. *Electric Machines and Power Systems*. 1979;**3**:341–355

[19] Borcherts, R.H., & Davis, L.C. 'Lift and drag forces for the attractive electromagnetic suspension systems'. *IEEE Transactions on Magnetics*. 1974;**10**(3):425–428

[20] Katz, R.M., Eastham, A., & Dawson, G. 'Integrated magnetic suspension and propulsion of guided ground transportation vehicles with a SLIM'. *IEEE Transactions on Magnetics*. 1979;**15**(6):1437–1439

[21] Eastham, J.F., & Rodger, D. 'The performance of induction levitators'. *IEEE Transactions on Magnetics*. 1984;**20**(5):1684–1686

[22] Bird, J., & Lipo, T. 'The experimental verification of the lift and guidance forces created by an electrodynamic wheel rotating over a split-sheet guideway'. *The 19th Internat. Conf. on Magnetically Levitated Systems and Linear Drives*; Dresden, September 2006 (Maglev2006.de)

[23] Bird, J., & Lipo, T.A. 'An electrodynamic wheel: an integrated propulsion and levitation machine'. *IEMDC'3*, Madison, Wisconsin, USA; 2003, pp. 1410–1416

[24] Simon, M.D., Heflinger, L.O., & Geim, A.K. 'Diamagnetically stabilized magnet levitation'. *American Journal of Physics*. 2001;**69**(6):702713

[25] Moon, F.C. 'Chaotic vibrations of a magnet near a superconductor'. *Physics Letters A*. 1988;**132**(5):249–251

[26] McMichael, C.K. Ma, K.B., Lamb, M.A., Lin, M.W., Chow, L., Meng, R.L., Hor, P.H., & Chu, W.K. 'Practical application in bulk superconducting magnet bearing applications'. *Applied Physics Letters*. 1992;**60**(15):1893–1895

Chapter 8
High-frequency electric suspension

Engineering is the professional art of applying science to the optimum conversion of natural resources to the benefit of man.

Ralph J. Smith

8.1 High-frequency stabilisation

Research and development programmes directed towards the evolution of micro-machines, such as electrostatic micromotors and fluidically driven micro-pumps, having dimensions in the sub-millimetre range when embedded within a planar substrate, have been growing rapidly and vigorously over the past 20 or so years. During the 1990s, major advances were made in the miniaturisation of electromechanical devices through the adoption of planar silicon technology as used in the fabrication of integrated circuits. This technology was commonly referred to as micro-electro-mechanical systems (MEMSs). At first sight, it may seem irrational to pursue energy efficiency in devices that are so small that the powers involved are measured in milliwatts but in fact as a proportion of useful output the losses at the MEMS scale can be massive. In particular, the frictional problems associated with bearings, and the losses generated are much more severe than in their macro-scale counterparts. The frictional (in relative terms) force associated with a micro-sized moving part or micro-bearing can actually become more than the motive force developed by the machine [1–3], in which case the device is useless. Because of this gross inefficiency, the accumulated losses in millions of micro-machines, in millions of watches, for example is by no means insignificant, thus needlessly wasting an awful lot of energy. Energy is predicted to become increasingly scarce this century, so even micro-device efficiency may be critical to the maintenance of an advanced civilisation.

A proposed solution to this problem has been sought by some researchers through the agency of electric field suspension, but in a manner quite distinct from electrostatic suspension addressed in Chapter 5. The idea was to use an AC system with the capacitor, forming the suspension element, embedded within a resonant LC circuit. In this case, the suspension fields cannot be electrostatic because by Maxwell's fourth equation (Chapter 2), a changing electric field induces a magnetic field. By 1992, it was reported [4] that stable suspension of a small conducting rotor within an electrostatic micromotor could be achieved when the rotor formed part of the

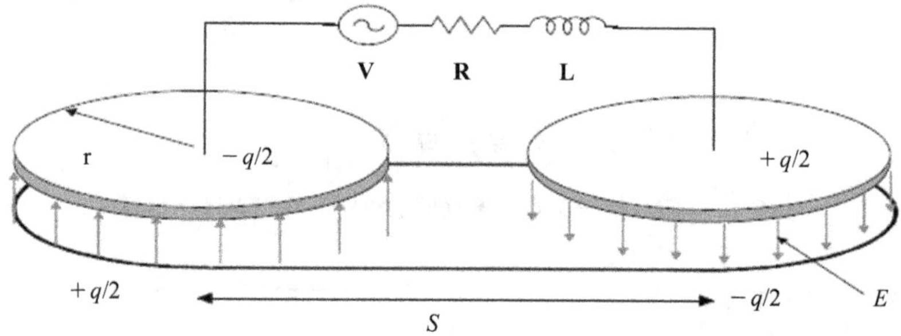

Figure 8.1 Schematic of lumped circuit unbalanced levitation scheme

capacitive element of a detuned resonant LC circuit. The published literature soon reported [5–7] that suspension had been demonstrated, in principle, for a small-scale electrostatic device, although admittedly for conducting pads that were still relatively large (area ~2 cm^2). These dimensions permit the easy realisation of the capacitive and inductive components of the resonant circuit at frequencies in the MHz range. The basic circuit arrangement underlying these developments is shown schematically in Figure 8.1.

8.1.1 Detuned LC circuit stabilisation

In our analysis and discussion of electromagnetic levitation forces, thus far, we have chosen to employ energy forms, from which the forces are derived, expressed in terms of basic electric and magnetic fields quantities. However, mathematically this is often unnecessarily complicated, since it so happens that convenient energy forms, expressed in terms of lumped circuit parameters, already exist. In low-frequency electrical engineering applications, where electromagnetic wavelength is very long compared to device size, this is usually the case, with electrical energy defined by lumped capacitance (C Farads) and magnetic energy by lumped inductance (L Henries). Electrical circuit analysis, including transmission line theory, actually represents a limited form of Maxwell's equations that apply strictly to lumped models of electrical devices.

In Chapter 2, it is shown that for a pair of flat parallel metallic electrodes connected to a DC source, the electric field is determined by the Gauss law (2.13). This field is related to the charge on the electrodes and the area of the float, and for the simple parallel plate system its charge storage capacity C is given by

$$C = \frac{\varepsilon_0 A}{t} \quad \text{(Farads)} \tag{8.1}$$

where A (m^2) is the area of the float, t (m) is the distance between the float and the fixed electrode and ε_0 is the permittivity of air. In terms of this lumped capacitance, the energy stored in the system is (see (3.17))

$$W_e = \frac{1}{2}CV^2 \quad \text{(Joules)} \tag{8.2}$$

Equally ubiquitous to the capacitor, in electrical circuits, is the lumped inductor that is usually represented by a conducting wire coiled around a soft iron core. In Appendix I, it is demonstrated that the inductance L of a coil can be determined, approximately, from the integral forms of Ampere's law and Faraday's law. It is defined in terms of magnetic flux linkages and the current in the coil (A2.16), and this formulation leads to the circuit version of Faraday's law, namely

$$V = L\frac{di}{dt} \tag{8.3}$$

The energy stored in an inductor can be readily calculated by noting that

$$W_m = \int_0^T Vi\,dt \quad \text{(Joules)} \tag{8.4}$$

where T is the time to build up the current i from zero to I. Combining (8.3) and (8.4) gives following the useful energy relation:

$$W_m = \frac{1}{2}LI^2 \quad \text{(Joules)} \tag{8.5}$$

where I is the final current in the coil.

The circuit relationship for resistive elements is expressed through Ohm's law, namely

$$V = IR \tag{8.6}$$

This is demonstrated in Appendix I.

With the basic circuit elements L, C and R now defined, we can proceed to the examination of an ingenious electrostatic levitation system [4,5] for micro-actuators, which is claimed to be unconditionally stable. It is shown schematically in unbalanced form in Figure 8.1 and in a balanced form in Figure 8.2. It comprises a conducting float (levitated plate) suspended between two (unbalanced) or four (balanced) electrodes connected to an alternating current source V, such that the instantaneous electric field enveloping the float produces levitation. The upper and lower electrode pairs are parallel connected to the source through a series connected inductor L and resistor R. For a float of mass m, an equilibrium suspension condition will occur when the difference between the upward electric force just balances the gravitational force mg. The electric force depends on charge separation in the plate, with negative free charge moving towards the positive electrode leaving exposed positive charge under the negative electrode. However, if the source is a simple DC battery this equilibrium position will be very unstable as was demonstrated in Chapter 3. It is clear from (3.27) and (3.28) that to achieve stabilisation in the electrostatic suspension arrangement depicted in Figure 3.13, the voltage V needs to increase as the gap gets wider and needs to decrease when the float moves closer to the fixed electrodes. In a device that is not too small, maintenance of the equilibrium position is not unduly difficult by sensing the float position and feeding the information to a closed loop control circuit designed to generate a voltage, and hence electrostatic

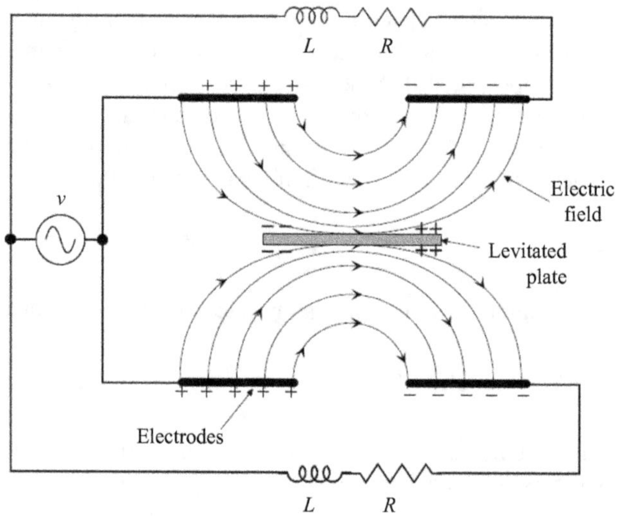

Figure 8.2 Schematic of lumped circuit balanced levitation scheme

forces, phased to counteract the float movement. However, stabilisation in this case is conditional on the accuracy to which the float position can be measured (not easy in MEMS scale devices) and on the control system sensitivity settings. It would obviously be preferable if the voltage adjustment were to occur automatically, as a result of float displacement. The system would then be deemed to be unconditionally stable, at least in the vertical direction. The scheme proposed by Kumar et al. achieves this by embedding the levitation system within a resonant circuit [4].

In Chapter 3, it is shown that the force supporting the float in Figure 8.1 is given by (3.25)

$$F_z = \frac{1}{2}\frac{\partial C}{dz}V_c^2 \tag{8.7}$$

where V_c is the capacitor voltage and the z-direction is normal to the float surface at $z = 0$. In an AC circuit driven by a sinusoidal source with frequency ω_s rad/s, the voltage across the capacitor formed by the fixed electrodes and the float can be expressed as

$$V_c = V_m \sin(\omega_s t + \varphi)$$

where V_m is the peak voltage and ϕ is a fixed phase dependent on R, L and C. Hence from (8.7) we get

$$F_z = \frac{1}{4}\frac{\partial C}{\partial z}V_m^2 - \frac{1}{4}\frac{\partial C}{\partial z}V_m^2\cos(2\omega_s t + 2\varphi) \tag{8.8}$$

If the source frequency ω_s is high enough, the inertia of the stator will ensure that the second term in this force equation is insignificant. The force will be influenced only by the first term with F_z increasing as C decreases, and vice versa for growing C, as required for stable equilibrium. It is assumed in forming this equation that the distance S (Figure 8.1) is small enough to consider that charge separation in the float is essentially instantaneous. For MEMS devices with very small suspended rotors, the operating frequency has to be pushed ever higher to ensure that the second force term in (8.8) remains insignificant. This, as we shall see, presents a considerable difficulty for lumped circuit designs.

8.1.2 Circuit resonance

For an R, L, C circuit fed from an AC source, theory indicates that the capacitor voltage is dependent on the relationship between the source frequency and the resonant frequency of the RLC circuit. This can be shown by applying a routine circuit analysis of the series RLC circuit. For such a circuit, the series impedance is the sum of the individual contributions from the circuit elements. That is

$$Z_s = R + j\omega_s L + \frac{1}{j\omega_s C} \tag{8.9}$$

Hence from Kirchoff's law the source voltage V is given by

$$V = Z_s i \tag{8.10}$$

where i is the instantaneous circuit current. This current also passes through the capacitor. Therefore

$$V_C = Z_C i = \frac{i}{j\omega_s C} \tag{8.11}$$

Consequently, by dividing (8.11) by (8.10), we can write

$$\frac{V_C}{V} = \frac{1}{j\omega_s Z_s C} = \frac{1}{(1 - \omega_s^2 LC) + j\omega_s RC}$$

It is not difficult, with some complex number manipulation, to demonstrate that in magnitude

$$\left|\frac{V_C}{V}\right| = \frac{1}{\left[(1 - \omega_s^2 LC)^2 + \omega_s^2 R^2 C^2\right]^{\frac{1}{2}}} \tag{8.12}$$

By using (8.12), the capacitor voltage in the series RLC circuit, as a proportion of the source voltage, is plotted as a function of a normalised drive frequency ($\omega_s\sqrt{LC_0}$) in Figure 8.3. C_0 is the capacitance of the levitation system for the float in its desired equilibrium position. R and L have been chosen such that

$$\frac{C_0 R^2}{L} = 0.08$$

The solid dark curve depicts the resonant profile of the voltage across the float–capacitor system when the levitated plate is located in the equilibrium position.

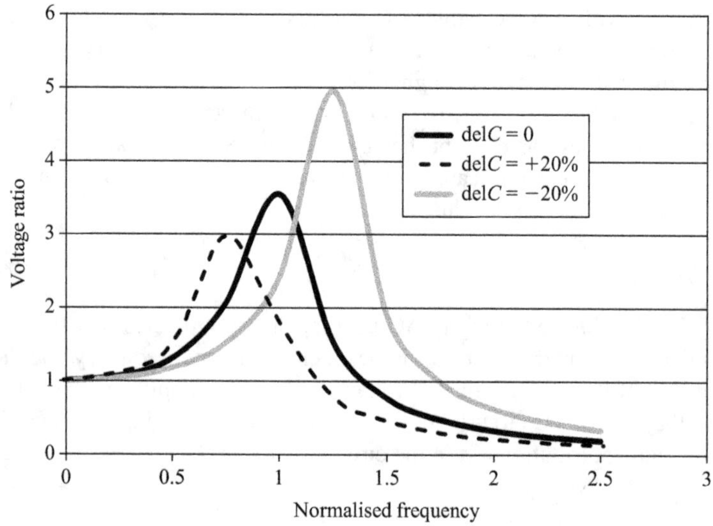

Figure 8.3 Capacitor voltage in series RLC circuit as a function of normalised drive frequency $\omega_s\sqrt{LC_0}$, where C_0 is the capacitance of the system at the desired float position

Consequently, the voltage peak occurs at $\omega_s\sqrt{LC_0} = 1$. If the plate is displaced upwards from the equilibrium position, the capacitance of the levitation system will rise. If the upwards movement is sufficient to procure a 20% increase in capacitance (relative to C_0), the dotted curve is produced that peaks at $\omega_s\sqrt{LC_0} = 0.75$. When the float drops away from the equilibrium position lowering the capacitance by 20%, it results in solid lightly shaded curve that maximises at $\omega_s\sqrt{LC_0} = 1.25$.

The significance to float stability of this resonant circuit behaviour is as follows. If R, L and C are selected to produce a resonant frequency at (say) ω_0, then peak capacitor voltage will be secured when $\omega_s = \omega_0$. At fixed frequency ω_s, any change in float position will result in a fall in C and a resultant diminution of the float levitation force. However, if ω_s is set slightly higher than ω_0, upwards displacement of the float produces a movement in the resonant curve to the left and the operating point at ω_s must slide down to a lower value of V_c. The levitation force thus declines (see (8.7)) and the plate returns to the equilibrium position under the force of gravity. On the other hand, a displacement downwards of the float causes the resonance curve to move to the right forcing the operating point to a higher voltage. The levitation force will increase in sympathy, lifting the plate back to its equilibrium position. The voltage changes and the concomitant forces are thus stabilizing. Evidently, the requirement for stability is that the resonant frequency of the series circuit at the equilibrium position (ω_0) should be less than the source frequency ω_s. In short, the stability criterion is

$$\frac{\omega_0}{\omega_s} < 1 \tag{8.13}$$

8.1.3 Float dynamics

The electric suspension arrangement described above is actually the analogue of a well-established magnetic levitation system developed for inertial bearings [2,3]. By reversing the roles of the inductor and capacitor in an open loop, unconditionally stable, magnetic suspension system is realised. That the electric levitation system is much less well established than magnetic systems is related to the fact that the former approach is restricted to devices of small size. More particularly there is an upper limit to the float–electrode gap width. This can be demonstrated by considering simple levitation of a flat float of area A and thickness t suspended below a single fixed electrode of the same area. The peak voltage required to levitate such a plate against the force of gravity, at a distance z from the stator plate, can be estimated using (3.26)

$$\frac{1}{4}\frac{\varepsilon_0 A V^2}{z^2} = mg = \rho A t g \tag{8.14}$$

The additional 'half' in the first term comes from replacing DC voltage by peak AC voltage, while ρ is the density of the material forming the float. Hence the peak voltage expressed in terms of the gap distance z is given by

$$V = \left[\frac{4\rho g}{\varepsilon_0}\right]^{1/2} (t)^{1/2} z \quad \text{(Volt)} \tag{8.15}$$

This equation is plotted in Figure 8.4, where the voltage in the levitation gap is compared with the Paschen breakdown voltage for an air gap between two flat charged electrodes. Two float options are plotted: one with thickness t of the copper plate equal to 2 mm and the second with a much thinner 0.02 mm (20 μm) plate. The thicker, heavier, plate obviously requires more voltage, than a thin plate, to

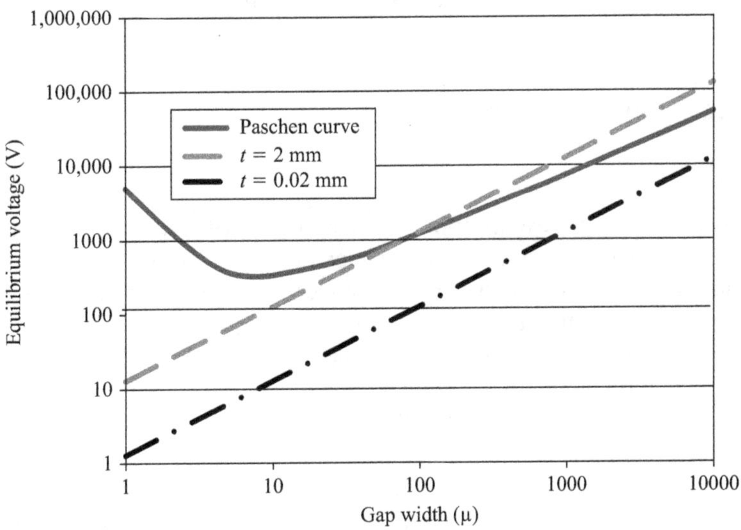

Figure 8.4 Operating limits for capacitive levitation system

maintain equilibrium at a chosen gap distance d. Eventually, at a separation of more than 100 μm, the voltage exceeds the Paschen limit for breakdown. For this example, safe operation requires a float that is equal to or less than about 200 μm in thickness.

The Paschen criterion dictates that levitation devices, which employ a resonated electrostatic capacitor for stability, must be small enough to avoid suspension voltage levels associated with air breakdown. But in addition, there is also a limit to how far miniaturisation can proceed. This is because of the necessity, in so doing, to move to higher and higher resonant frequencies. This boundary is reached when the free space wavelength at the source frequency is similar in magnitude to the dimensions of the series circuit elements and associated wire connections. At this point, high losses due to radiation are incurred with unacceptable levels of resonance damping. The usual solution to this radiation problem is to replace the circuit resonator by a cavity resonator.

In general, attempts at circuit component miniaturisation, consistent with achieving a levitation device that is in sympathy with integrated circuit dimensions as in MEMS, have been fraught with difficulty, perhaps not too surprisingly. Capacitance values in the levitator are very low, of the order of 10 pF, while integrable inductors are in the nanoHenry range. This combination tends to push resonant frequencies above 1 GHz. It then becomes a 'losing battle' because lumped element resonators are increasingly difficult to realise much above 1 GHz because of stray electromagnetic fields, radiation and uncontrolled interactions and couplings between components. The eventual outcome is usually the abandonment of *L–C* circuits in favour of microwave (10 GHz approximately) cavity resonators, which are much more amenable to miniaturisation.

As will be discussed in section 8.3, recent research has shown that suitable field patterns, not unlike those in Figure 8.2, can be set up using metal cavities [8,9]. It is demonstrated there that a 'floating' wall can be suspended by the leakage electric fields of a cavity resonator without quenching the resonance.

8.2 Electromagnetic pressure in cavity resonator

8.2.1 TE_{101} mode resonator

Before we proceed to the examination of electric field levitation, which uses a cavity to provide stable suspension instead of a lumped series resonant circuit as described in section 8.1, we need to spend some time reviewing the nature of electromagnetic forces and pressure within pill-box resonators. In Chapter 2, we observed that the solution to Maxwell's equations in a rectangular waveguide has the form

$$\begin{aligned} H_z &= H_0 \cos\frac{\pi x}{a}\exp(-j(\bar{\beta}_{10}z-\omega t)) \\ H_x &= j\frac{\bar{\beta}_{10}aH_0}{\pi}\sin\frac{\pi x}{a}\exp(-j(\bar{\beta}_{10}z-\omega t)) \\ E_y &= -j\frac{kaH_0}{\pi}\eta_0\sin\frac{\pi x}{a}\exp(-j(\bar{\beta}_{10}z-\omega t)) \end{aligned} \tag{8.16}$$

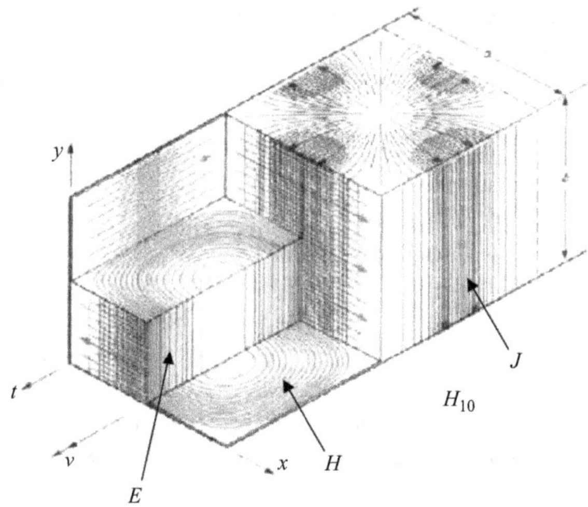

Figure 8.5 TE$_{10}$ mode – field and current pattern

Here H_0 is the magnitude of the axial field component of the mode, while all other parameters are defined in Chapter 2. The electric (E-annotated traces) and magnetic (H-annotated traces) field distributions for this mode, based on (8.16), are depicted in Figure 8.5. The J-annotated traces represent surface current density.

The forces that we are interested in exploring here are those due to the modal electric field. In this case, it is convenient to adjust (8.16) so that E_y becomes the reference field. If the magnitude of this field is $E_0 = \eta_0 H_0$, then (8.16) can be recast in the following form

$$\begin{aligned} H_z &= \frac{j\pi}{\omega\mu_0 a} E_0 \cos\frac{\pi x}{a} \exp(-j(\bar{\beta}_{10}z - \omega t)) \\ H_x &= j\frac{\bar{\beta}_{10}E_0}{\omega\mu_0} \sin\frac{\pi x}{a} \exp(-j(\bar{\beta}_{10}z - \omega t)) \\ E_y &= E_0 \sin\frac{\pi x}{a} \exp(-j(\bar{\beta}_{10}z - \omega t)) \end{aligned} \tag{8.17}$$

In phasor notation, the electric field E_y can be written as

$$E_y = \frac{E_0}{2} \sin\frac{\pi x}{a} \exp(-j\bar{\beta}_{10}z) \tag{8.18}$$

Equation (8.18) represents a 'forward' wave travelling in the $+z$-direction within the guide. The division by two is introduced for mathematical convenience. To form a cavity resonator we first need to terminate the guide in a perfect shorting wall at $z = 0$ (say). At this wall the boundary conditions demand that $E_{\tan}$ at the wall is everywhere zero. This is possible only if the wave in the guide comprises a forward and an equal backward (reflection) component. That is

$$\begin{aligned} E_{\mathrm{T}} &= \frac{E_0}{2}\sin\frac{\pi x}{a}\exp(-j\bar{\beta}_{10}z) - \frac{E_0}{2}\sin\frac{\pi x}{a}\exp(+j\bar{\beta}_{10}z) \\ &= E_0\sin\frac{\pi x}{a}\sin\bar{\beta}_{10}z \end{aligned} \tag{8.19}$$

The mathematical form of this equation is representative of a standing wave. In addition to being equal to zero at $z = 0$, as required by the imposed boundary condition, E_{T} is also equal to zero at

$$\bar{\beta}_{10}z = n\pi \tag{8.20}$$

This equation implies that a secondary shorting wall can be introduced, without upsetting the standing wave pattern, at distance L from the $z = 0$ termination, if

$$L = \frac{n\pi}{\bar{\beta}_{10}} \tag{8.21}$$

When $n = 1$, the shorting walls at $z = 0$ and $z = L$ trap a portion of the standing wave pattern referred to as the TE_{101} rectangular cavity mode.

The field distribution for the TE_{101} mode can be expressed in the form

$$\mathbf{E} = \hat{\mathbf{a}}_{\mathrm{y}}E_0\sin\frac{\pi x}{a}\sin\frac{\pi z}{L} \tag{8.22}$$

$$\mathbf{H} = \hat{\mathbf{a}}_{\mathrm{x}}\frac{j\bar{\beta}_{10}}{\omega\mu_0}E_0\sin\frac{\pi x}{a}\cos\frac{\pi z}{L} + \hat{\mathbf{a}}_{z}\frac{j\pi}{\omega\mu_0 a}E_0\cos\frac{\pi x}{a}\sin\frac{\pi z}{L} \tag{8.23}$$

where $\hat{\mathbf{a}}_{\mathbf{x}}$, $\hat{\mathbf{a}}_{\mathbf{y}}$ and $\hat{\mathbf{a}}_{\mathbf{z}}$ are unit vectors and β_{10} is the propagation coefficient for the z-directed waveguide that forms the cavity. For a cavity with perfectly conducting walls such a mode, once set up, would 'ring' forever, in much the same way as a frictionless pendulum would oscillate forever once started. However, in practice cavity walls are never perfectly conducting and a small amount microwave power, usually inserted by means of a coaxial probe, is required to maintain the resonance.

8.2.2 *Forces on wall of cavity*

For the cavity depicted in Figure 8.6, let us consider the forces acting on a 'free' plate, of width w and length ℓ, located in a centrally placed aperture of similar dimensions. These can be estimated as shown below. It is assumed that for small apertures, the TE_{101} fields are undisturbed by the aperture, and also that a mechanism exists for allowing charge to build up on the 'free' plate (cf. Figure 8.2).

If the time average force acting on the float is F_{e}, then the work done in moving it a distance dy in the y-direction is $W_{\mathrm{e}} = F_{\mathrm{e}}\,\mathrm{d}y$. But W_{e} must equal the change in energy in the cavity caused by this movement. Given that the energy density in the cavity electric field (*peak value*) is given by (see (3.21))

$$U_{\mathrm{e}} = \frac{\varepsilon_0|E^2|}{4} \tag{8.24}$$

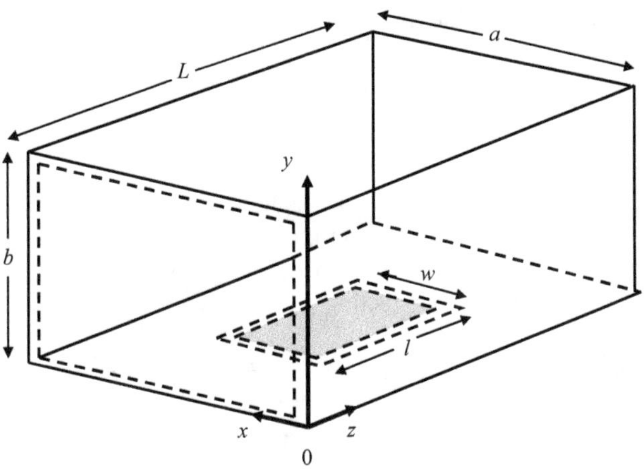

Figure 8.6 Rectangular waveguide cavity with float forming a portion of the lower wall

then the time averaged work done on the plate is given by

$$F_e \mathrm{d}y = \int_{a/2-w/2}^{a/2+w/2} \int_{L/2-\ell/2}^{L/2+\ell/2} \frac{\varepsilon_0 |E^2|}{4} \mathrm{d}x\, \mathrm{d}z\, \mathrm{d}y \tag{8.25}$$

Evidently, the time averaged 'attractive' electric force (F_e) becomes

$$\begin{aligned} F_e &= \frac{\varepsilon_0 E_0^2}{4} \iint \sin^2 \frac{\pi x}{a} \sin^2 \frac{\pi z}{L} \mathrm{d}x\, \mathrm{d}z \\ &= \frac{\varepsilon_0 E_0^2}{4} \left[\left(\frac{w}{2} + \frac{a}{2\pi} \sin \frac{\pi w}{a} \right) \left(\frac{\ell}{2} + \frac{L}{2\pi} \sin \frac{\pi \ell}{L} \right) \right] \quad \text{(Newtons)} \end{aligned} \tag{8.26}$$

Similarly, the time averaged 'repulsive' Lorentz force can be estimated from the following equation:

$$F_m = \int_{a/2-w/2}^{a/2+w/2} \int_{L/2-\ell/2}^{L/2+\ell/2} \frac{\mu_0 |H^2|}{4} \mathrm{d}x\, \mathrm{d}z \tag{8.27}$$

On applying (8.14) this gives

$$\begin{aligned} F_m &= -\frac{\mu_0}{4} \left(\frac{\beta E_0}{\omega \mu_0} \right)^2 \iint \sin^2 \frac{\pi x}{a} \cos^2 \frac{\pi z}{L} \mathrm{d}x\, \mathrm{d}z \\ &\quad - \frac{\mu_0}{2} \left(\frac{\pi E_0}{\omega \mu_0 a} \right) \iint \cos^2 \frac{\pi x}{a} \sin^2 \frac{\pi z}{L} \mathrm{d}x\, \mathrm{d}z \\ &= -\frac{E_0^2}{4\omega^2 \mu_0} \left\{ \begin{array}{l} \beta^2 \left(\frac{w}{2} + \frac{a}{2\pi} \sin \frac{\pi w}{a} \right) \left(\frac{\ell}{2} - \frac{L}{2\pi} \sin \frac{\pi \ell}{L} \right) \\ \quad + \frac{\pi^2}{a^2} \left(\frac{w}{2} - \frac{a}{2\pi} \sin \frac{\pi w}{a} \right) \left(\frac{\ell}{2} + \frac{L}{2\pi} \sin \frac{\pi \ell}{L} \right) \end{array} \right\} \end{aligned} \tag{8.28}$$

Combining (8.26) and (8.28) yields

$$\left|\frac{F_e}{F_m}\right| = \frac{k_0^2}{\beta^2\left(\dfrac{1-\frac{L}{\pi\ell}\sin\frac{\pi\ell}{L}}{1+\frac{L}{\pi\ell}\sin\frac{\pi\ell}{L}}\right)+\dfrac{\pi^2}{a^2}\left(\dfrac{1-\frac{a}{\pi w}\sin\frac{\pi w}{a}}{1+\frac{a}{\pi w}\sin\frac{\pi w}{a}}\right)} \tag{8.29}$$

This relationship is shown graphically in Figure 8.7 for a rectangular waveguide cavity with $a = 2.0$ cm and $L = 2.265$ cm, operating at 10 GHz. It shows that for small apertures, for which w/a and ℓ/L are less than about 0.2, the levitation force swamps the Lorentz force on a conducting plate located within the aperture.

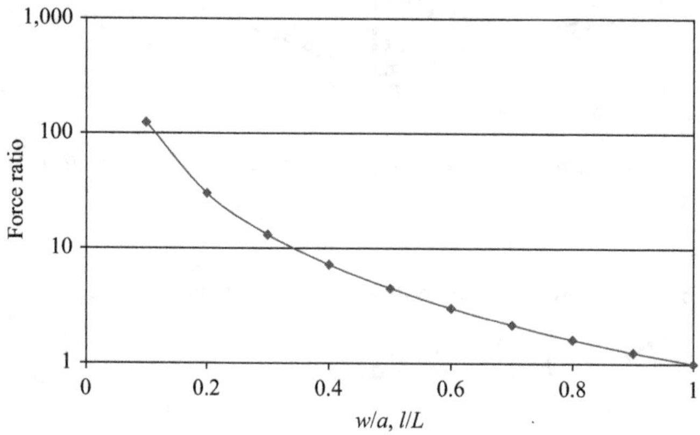

Figure 8.7 Force ratio as a function of w/a $(=\ell/L)$*:* $a = 2.0$ *cm,* $L = 2.265$ *cm,* $\lambda = 3.0$ *cm*

However, this force ratio enhancement in favour of the E-field, achieved by means of judicious location of the aperture within the modal field pattern, is gained at the expense of available levitation force, which reduces approximately in proportion to the aperture area for small apertures (8.26). This can be quantified as follows. The energy stored within a resonant cavity is related to the Q of the cavity, and P the power delivered to it, through the equation [10]:

$$W_g = \frac{PQ}{2\pi f_0} \quad \text{(Joules)} \tag{8.30}$$

where f_0 is the resonant frequency. Also for a rectangular waveguide cavity the energy stored in the TE_{101} field pattern is given by [10]

$$W_g = \frac{\mu_0}{2}\left(\frac{\pi}{\omega\mu_0 a}\right)^2 E_0^2 abL\left[\frac{a^2}{L^2}+1\right] \quad \text{(Joules)} \tag{8.31}$$

Low-loss rectangular cavities operating in the TE_{101} mode display Q-values of the order of 10,000. Consequently, for a 2 cm wide and 2.265 cm long cavity, resonating at 10 GHz, an input power level 1 W would result in an energy storage level of 2 μJ.

Using (8.31), this equates to a maximum electric field E_0 of 3.15 kV/cm at the centre of the aperture plate. Hence, the following relationship (using (8.26)) between the electric field force on a conducting plate positioned within a centrally located aperture, of width w and length ℓ, can be formed

$$F_e = 0.294 \times 10^{-4}\left[\left(\frac{w}{a} + \frac{1}{\pi}\sin\frac{\pi w}{a}\right)\left(\frac{\ell}{L} + \frac{1}{\pi}\sin\frac{\pi \ell}{L}\right)\right] \quad (8.32)$$

The magnitude of this electric field force F_e in μN is plotted in Figure 8.8. For small aperture sizes ($w/a < 0.25$), the force increases approximately in proportion to the area of the aperture. It maximises at a value of 25 μN when the aperture covers the entire broad wall. The F(lev) force plotted in Figure 8.8 represents the net force experienced by the aperture plate when the opposing Lorentz force is introduced. When w/a and ℓ/L are increased in magnitude beyond about 0.6, the force available for levitating the aperture float rapidly diminishes towards zero, as it should do for the TE_{101} mode, when the aperture plate becomes equal to the cavity wall in area. There is no photon pressure on the top and bottom walls of a rectangular waveguide supporting the TE_{10} mode since it is formed from TEM waves reflecting off the sidewalls only (see Chapter 2).

It is clear from the above, that for effective electric field levitation of a float that forms part of a cavity wall, the float must be immersed in the electric field and located in a position where it is not significantly influenced by the magnetic field.

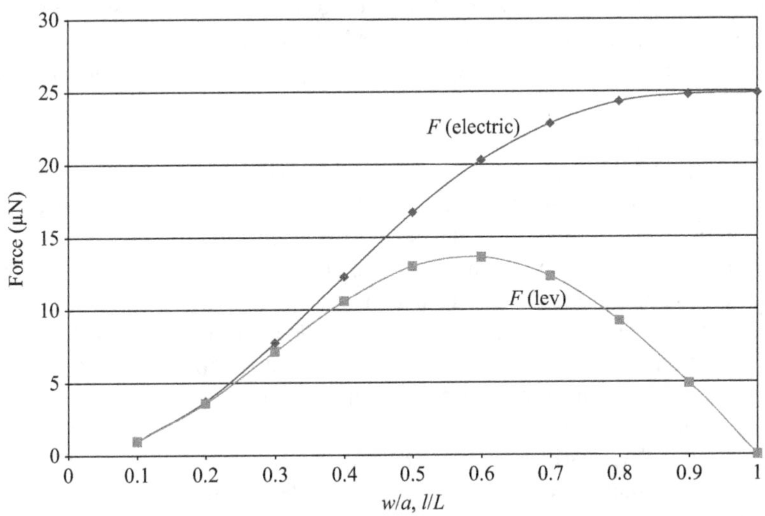

Figure 8.8 Force in μNewtons on a rectangular conducting plate in a centrally located aperture in the broadwall of a rectangular waveguide cavity supporting the TE_{101} mode ($a = 2$ cm, $L = 2.265$ cm, $f_0 = 10$ GHz, $Q = 10{,}000$, $P = 1$ W)

8.2.3 Application to micro-actuators

To assess whether the forces predicted by the above equations have significance for levitation in micro-actuators, it has been estimated that 15 μN is the force required to levitate a 1.45 mg nickel rotor of an electrostatically driven micromotor [11]. At the state-of-the-art in millimetre-wave technology that pertained in around the year 2000, the suggested power level of 1 W was rather too high for generation by a source that could be considered sufficiently compact for integration into a micro-actuator. However, forces in the 15 μN range could be achieved with much less power by employing a re-entrant cavity configuration and so the research in the mid-1990s proceeded in this direction. Re-entrant cavities can readily provide the necessary separation of the electric force from the Lorentz force while, at the same time, concentrating the electric field at the float. Cavities of this description, and in particular their levitation potential, are assessed in the next section.

8.3 Evolution of re-entrant cavity levitator

Re-entrant cavities have been familiar to electronic engineers, particularly those with an interest in high-power vacuum tubes for radar, for some considerable time [12–15]. This early interest was mainly concerned with the creation of cavities with small interaction gaps and high electric field strengths in order to modulate the velocity of electrons in a focused beam traversing the gap. It was well understood by radar scientists during the Second World War that the introduction of a square or cylindrical post into a pill-box cavity (as suggested in Figure 8.8) produced a significant reduction in the cavity resonant frequency. Hence, given that for a resonant circuit

$$f_{\text{res}} = \frac{1}{2\pi\sqrt{LC}}$$

it was not difficult to surmise that this must be due to enhanced capacitive effect in the cavity. This of course is associated with electric field concentration in the gap between the free end of the post and the cavity wall (see Figure 8.9). A tabulated set of graphs termed the Sperry curves [12], with improvements suggested in reference 16, can be very useful for determining the major dimensions of a resonant cavity of this type, when it is required to operate at a specified resonant frequency and with prescribed gap dimensions. Even in these days of sophisticated electromagnetic simulation packages, which can generate solutions to most boundary value problems, the software still requires to be furnished with realistic initial parameter values at the commencement of a complex cavity design, if rapid convergence on a valid solution is to be secured. The Sperry curves can provide this data and so these long-established design tools continue to retain significant value in re-entrant cavity design.

The relevance of the above to levitation systems is indicated in Figure 8.9. It shows a pair of identical re-entrant cavities in close proximity, with leakage apertures on the lower face, each directly below their respective re-entrant post. The cavities

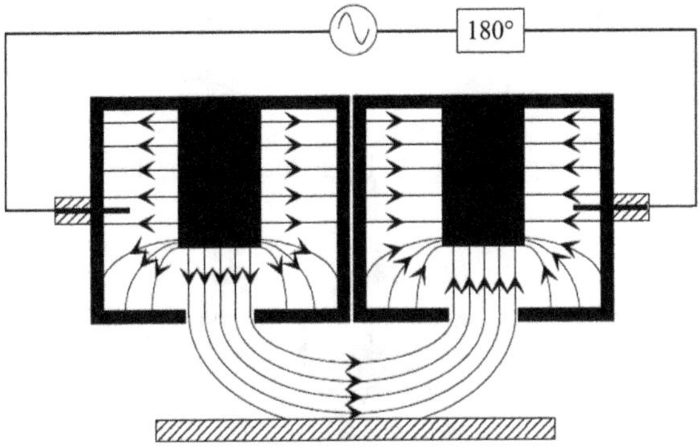

Figure 8.9 *Schematic drawing of coupled re-entrant cavity system with leakage aperture separated by distance S*

are shown as fed, through a phase shifter, from a single microwave source and excited by standard coaxial probes. If the phase shifter is set at 180, so that the fields in one cavity are the mirror image of those in the other, a leakage field distribution is set up, between the cavities and the float, which is the equivalent of that shown in Figure 8.2. Consequently, if the fields are strong enough, stable suspension of the float is predicted, in much the same manner as described in section 8.1.

Simple geometrical requirements dictate that the float is suspended below the lower wall of the cavity pair, and not as suggested by Figure 8.6 within the apertures. This raises the problem of possibly deleterious field magnitude reduction as the fields pass from the re-entrant cavity gap through an aperture to the space above the float. However, it turns out that this field attenuation due to the aperture is not an issue [9] provided the plate is close enough to the cavity to be sensibly immersed in the near fields. In practice, it is very unlikely that this would not be the case.

8.3.1 Figure of merit for cavity resonator

If the levitation plate in the vicinity of the cavity gap can be considered to reside in purely capacitive fields, as indicated in Figure 8.9, a basic criterion for the presence of effective electric field levitation can easily be established. Let the instantaneous peak electric field in the gap be E V/m and the instantaneous charge on each 'plate' be $q/2$ coulombs. For a small float separation from the cavity structure, the electric field in the gap will be essentially uniform as depicted schematically in Figure 8.1. Therefore, the electric field force F_e on the movable plate is

$$F_e = \frac{qE}{2} \quad \text{(Newtons)} \tag{8.33}$$

If the cross-sectional area of each gap is $A/2$ square-metres and the charge density on each plate is $\sigma = Aq$ C/m^2, then (8.1) becomes

$$F_e = \frac{\sigma AE}{2} = \frac{\varepsilon E^2}{2}A \quad \text{(Newtons)} \tag{8.34}$$

For r.f. fields this becomes

$$F_e = \frac{\varepsilon E^2}{4} A \quad \text{(Newtons)} \tag{8.35}$$

The magnetic field in the gap region can be deduced from the Faraday law, namely

$$\text{Circ } \mathbf{H} = -\frac{\partial}{\partial t} \text{flux } \mathbf{D} \tag{8.36}$$

Hence, for a cylindrical gap volume, at radius r in the gap

$$\begin{aligned} 2\pi r H &= -j\omega D \pi r^2 \\ \therefore\ H &= \frac{-j\omega r D}{2} \qquad \text{(A/m)} \end{aligned} \tag{8.37}$$

The current flowing radially in the plate at a radius r from the central axis of the cylindrical gap volume can be defined in terms of the tangential magnetic field present, therefore

$$i_r = 2\pi r H \qquad \text{(Amperes)} \tag{8.38}$$

Consequently, the opposing magnetic force at radius r is

$$\begin{aligned} dF_m &= B i_r\, dr \\ &= \frac{j\omega\mu}{2} rD \cdot 2\pi r H\, dr \\ &= \frac{\omega^2 \varepsilon^2 \mu \pi}{2} E^2 r^3 dr \end{aligned} \tag{8.39}$$

The total magnetic force on the movable plate is obtained by integrating dF_m from 0 to R for each gap. That is

$$\begin{aligned} F_m &= \frac{\omega^2 \varepsilon^2 \mu \pi}{2} E^2 \int_0^R r^3 dr \\ &= \frac{k^2 \varepsilon}{32\pi} A^2 E^2 \quad \text{(Newtons)} \end{aligned} \tag{8.40}$$

where $k^2 = \omega^2 \mu\varepsilon$ and $A/2 = \pi R^2$. For AC fields and for two gaps this becomes

$$F_m = \frac{k^2 \varepsilon}{32\pi} A^2 E^2 \qquad \text{(Newtons)} \tag{8.41}$$

Hence, from (8.35) and (8.41) we have

$$\frac{F_e}{F_m} = \frac{2c^2}{\pi f^2 A} \tag{8.42}$$

where c is the speed of light and f is the operating frequency for the cavity system. The above derivation is based on the assumption that the gap electric fields are essentially 'electrostatic' in nature. For this to be the case propagation delays must be negligible, within and between the capacitor gaps. This requires $S \ll \lambda$. At 10 GHz, for example propagation effects will be small if S is equal to, or less, than 0.1λ, that is $S = 3$ mm. The maximum value of A, to optimise the levitation force, is therefore of the order of 14.1 mm^2 that results in a force ratio of

$$\frac{F_e}{F_m} = 40 \tag{8.43}$$

If the capacitor gap area A is reduced in order to improve this ratio, the available levitation force (largely F_e) becomes progressively weaker (8.35). Equation (8.43) thus represents a figure of merit for the lumped levitation system with two capacitive elements.

8.3.2 Levitation using detuned re-entrant cavity resonator

The advantage of employing, microwave or millimetre-wave, coupled cavity resonators rather than LC circuits in a detuned resonator stabilisation technique is that it can be applied to much smaller actuators and bearings. This section outlines the requirements for successful implementation of levitation systems using microwave and millimetre-wave resonant structures to provide lifting forces that are responsive to float position. The basic electric field distribution, required for levitation and depicted in Figure 8.2, can also be produced by a range of re-entrant two-cavity resonator geometries as shown earlier. In such cavities, the electric stored energy in the cavity is spatially separated from the region of magnetic energy storage. Consequently, these cavities can generate field distributions that are also capable of providing levitation forces on a free conducting plate immersed within them because the plate is well shielded from any magnetic (Lorentz) forces. One possible arrangement, which has been examined in detail elsewhere [8,9], is shown in Figure 8.9. It comprises a pair of re-entrant cavities with leakage apertures opposite to the base of the re-entrant cavity posts. It is assumed with this arrangement that the distance between the leakage apertures is short relative to the free space wavelength at the frequency of operation. This can be achieved by dielectric loading the cavities. As the figure shows, the 'levitating' conducting plate is located parallel to the base of the coupled cavities within the fringing fields of the leakage apertures.

Vertical movement of the plate within the fringing fields has the desired effect of detuning the frequency matched cavities that is a pre-requisite of securing a quasi-stable levitation force. This has been confirmed by network analyser measurements on a coupled cavity system as depicted in Figure 8.10. It was designed around Tekelec fine tuning screws having a diameter of 4.1 mm. The cavity dimensions were chosen

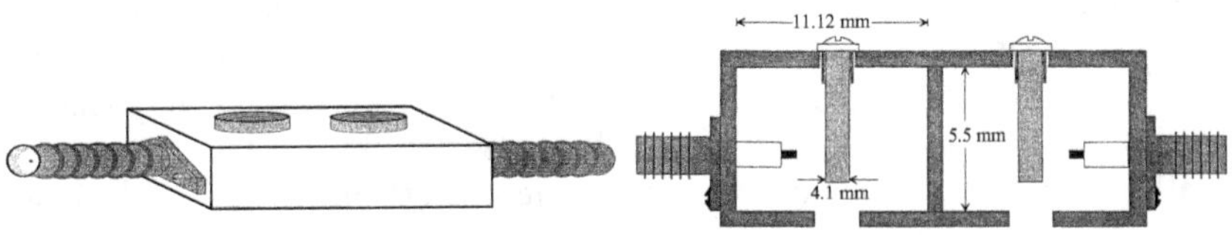

Figure 8.10 Re-entrant two-cavity system used in measurements (a) three-dimensional view, (b) dimensioned drawing

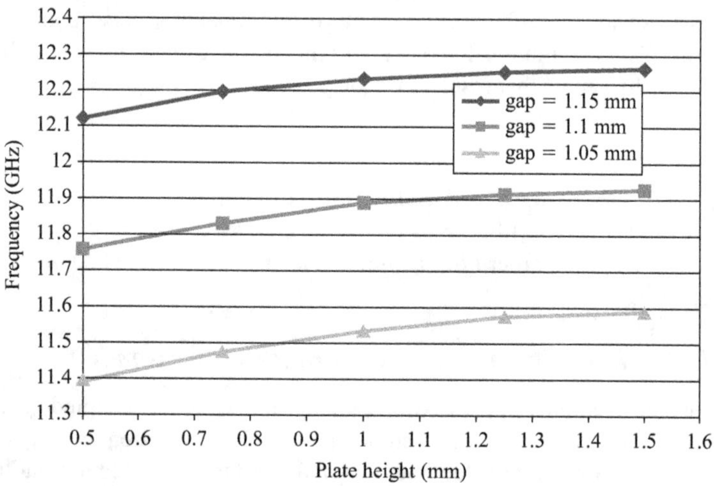

Figure 8.11 Measured resonant frequency versus plate height for the cavity levitation system at four re-entrant gap settings

to produce a basic resonant frequency for the system of 10.5 GHz in the absence of leakage apertures. Very precise control of the resonant frequency is provided by the Tekelec elements and Q-values in the 5,000–50,000 range, over the frequency range 10.0–12.0 GHz, were indicative of negligible leakage past the tuning screws. Typical results for resonant frequency values are presented as a function of levitation plate distance from the cavity base in Figure 8.11.

These show that on cavities with 1.5 mm diameter apertures the cavity resonant frequency changes approximately linearly with plate distance Δz from the base-plate of the cavities at a rate of 90 MHz/mm in the range 0.5 mm $<$ gap $<$ 1.5 mm.

A re-entrant cavity levitation system, as depicted in Figure 8.9, designed to operate at the same frequency as a simple TE_{101} mode rectangular cavity, has a volume that is ten times smaller than that of the rectangular cavity. Consequently, if it exhibits the same level of Q, and is fed with a similar power level, the energy density in the re-entrant cavity is ten times that of the equivalent TE_{101} mode cavity. However, this figure ignores the field enhancement in the re-entrant cavity gap. Since the electric field is largely confined to the re-entrant gap that is typically 30 times smaller than the volume of the cavity, the electric energy density in the vicinity of the leakage aperture is possibly as much as 300 times greater than is available in a simple rectangular cavity. Hence using Figure 8.8, it can be estimated that the levitation force, developed by a re-entrant cavity pair with the dimensions of Figure 8.10b, operating at 10 GHz, will be approximately 6 mN. It is assumed that for the cavities, Q is typically of the order of 10,000, the apertures are small (diameter $< 0.1\lambda = 3$ mm), and that the drive power is 1 W. Unfortunately, the electric field in the interior cavity gap exceeds the air breakdown level of 27 kV/cm – although this can be rectified by using evacuated cavities, with a sealing 'window' in the aperture.

The question that now needs to be asked is: "Can 6 mN suspend a float overlaying the leakage apertures?" To answer this question we need to know the gravitational force on such a float. For a float formed from an aluminium plate of thickness $t = 0.5$ mm (say), of length $a = 22$ mm and of width $b = 5$ mm (large enough to cover both apertures), we get

$$F_{\mathrm{g}} = \rho\, tabg \quad \text{(Newtons)}$$

where ρ is the density of aluminium and g is the gravitational acceleration. Hence we obtain, for the gravitation force on the plate

$$\begin{aligned} F_{\mathrm{g}} &= 2.7 \times 10^3 \times 0.5 \times 22 \times 5 \times 10^{-9} \times 9.807 \\ &= 1.46 \text{ mN} \end{aligned}$$

Given that the lifting force is proportional to the power delivered to the cavity, this result suggests that a drive power of 240 mW would be sufficient to levitate this plate with the two-cavity levitation system depicted in Figure 8.9, when operating at the rather low frequency of 10 GHz.

8.3.3 Force measurements

Calculations based on the 'universal Q-curve' for resonators [17], predicts that to produce, by detuning, a 10 dB suppression of the resonant fields, essentially quenching the cavity fields, we require

$$\Delta f = \frac{1.6 f_0}{Q} \quad \text{(Hz)} \tag{8.44}$$

For the values of resonant frequency and Q given above this equation gives a Δf of the order of 2.5 MHz. Therefore, to achieve quenching of the cavity resonance, at a fixed frequency, a plate movement of ~0.03 mm is required. The work done, in moving the levitation plate the distance required for quenching cavity oscillation, can be equated to the change in energy stored in the cavity system, and from (8.30), this generates the following approximate relation for the average levitation force:

$$F_{\mathrm{L}} \approx \frac{QP}{2\pi f_0 \Delta z} \tag{8.45}$$

where P is the microwave power supplied to the cavity and Δz is the distance moved by the plate. The estimated levitation force as a function of drive power is shown in Figure 8.12 with the gravitation force on the 1.45 mg micro-actuator rotor [18] shown for comparison. The theoretical estimate applies to a 2 mm × 2.3 mm rectangular aperture. The measured result indicates, that at a frequency in the microwave range (10–12 GHz), and with 1.5 mm diameter leakage apertures, drive power levels of the order of 4–5 mW are required by the two-cavity re-entrant system (Figure 8.9), to levitate a conducting plate with a weight of 1.6 mg. This is approximately the weight of the rotor of, for example the electrostatically driven micro-actuator cited earlier. This drive level could be lowered considerably by introducing larger leakage apertures into the system as the theoretical curve shows, with the proviso that the Lorentz force is kept small by comparison with the electric levitation force.

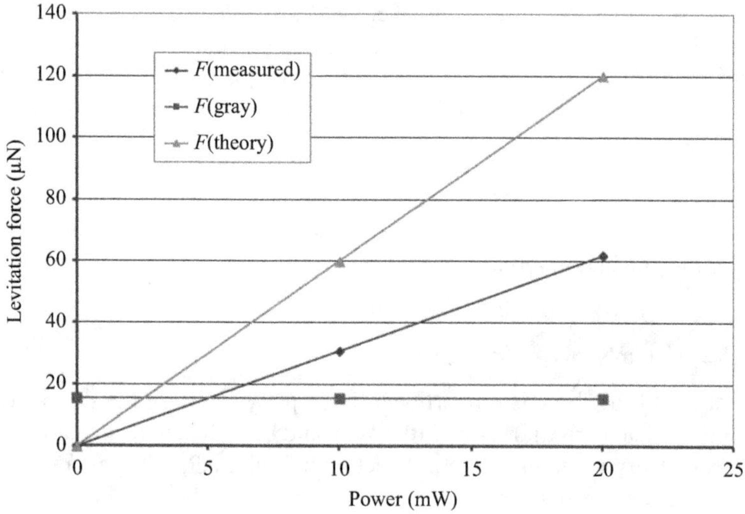

Figure 8.12 Estimated levitation force versus microwave power input, (F(measured) is based on resonant shift measurements for a 1.5 mm diameter aperture, F(theory) applies to a 2 mm × 2.3 mm rectangular aperture, F(grav) is the gravitational force on a 1.6 mg float)

8.4 Summary

This chapter has demonstrated, firstly, that by embedding an electric field levitation system within either a lumped circuit resonator, or within a re-entrant cavity resonator, it is possible to secure unconditionally stable levitation forces. The 'trick' required to achieve stability is to tune the resonators to a frequency that slightly exceeds the frequency of the power source. The degree of detuning depends very much on the Q of the resonator employed.

Secondly, by employing basic field relations for the dominant mode together with fundamental energy expressions for cavity resonators, the primary requirements that must be met by a cavity-based electric field levitation system have been established. When these requirements are fulfilled useful levitation of a 'free' conducting plate immersed, normal to, and within the cavity electric field pattern, is available. In particular, the criterion, which ensures that the electric field force associated with levitation outweighs the opposing Lorentz force, is quantified.

Thirdly, electrical forces that are sufficient to levitated micro-actuator sized rotors are shown to be attainable from an X-band (10–12 GHz) two-cavity re-entrant system with drive levels in the milliwatt range. It has also been demonstrated that cavity designs, which provide a significant degree of isolation between the electric field force and the Lorentz force, are essential, if successful implementation of cavity resonators in a levitation system is to be achieved.

Finally, resonant frequency and scattering parameter measurements on a re-entrant cavity levitation system comprising two adjacent cavities fed in anti-phase confirm that at X-band a few milliwatts of microwave power is sufficient to levitate micro-actuator sized rotors weighing typically between 1 and 2 mg.

In general terms, microwave cavities exhibit shapes that can readily be envisaged as forming frictionless bearings in electrical motors of generators. The stumbling block to progress, at present, is that the generation of power at microwave frequencies still remains relatively expensive, particularly at very high frequencies, in the millimetre wavelength bands, required to make cavities compatible with bearing sizes in small machines. But current trends suggest that this will not be the case for much longer.

8.5 References

[1] Gilinson Jr, P.J., Denhard, W.G., & Frazier, R.H. *A magnetic support for floated inertial instruments*. S.M. Fairchild Publication Fund paper no. FF-27. Institute of Aeronautical Sciences, New York, May 1960

[2] Kaplin, B.Z. 'Analysis of a method for magnetic levitation'. *Proceedings of IEE*. 1967;**114**:1801–1804

[3] Kaplan, B.Z., & Regev, D. 'Dynamic stabilization of tuned circuit levitators'. *IEEE Transactions*. 1976;**MAG-12**(5):556–559

[4] Kumar, S., Cho, D., & Carr, W.N. 'A proposal for electrically levitating Motors'. *Sensors and Actuators A*. 1990;**24**:141–149

[5] Kumar, S., Cho, D., & Carr, W.N. 'Experimental study of electric suspension for micro-bearings'. *Journal of Microelectromechanical Systems*. 1992;**1**(1):23–30

[6] He, G.H., Chen, K.S., Tan, S.S., & Wang, W.Y. 'Electrical levitation for micromotors and micro-accelerometers'. *Proceedings of the 8th International Conference on Solid State Sensors and Actuators*; Stockholm, 1995, pp. 442–445

[7] Toda, R., Takeda, N., Nakamura, S., & Esashi, M. 'Electrostatically levitated spherical 3-axis Accelerometer'. *15th International Conference on Electomechanical Systems*; Las Vegas, 2002, pp. 710–713

[8] Sangster, A.J., & Kelly, M. 'Electromagnetic levitation using the fringing fields of a microwave cavity'. *ANTEM '96, Conference Proceedings*; Montreal, Canada, 7–9th August 1996, pp. 99–102 (ISBN-0-9692563-5-3)

[9] Sangster, A.J., McErlean, E., Beale, G., Kelly, M., & Smith, P. 'Coupled reentrant cavity system for electromagnetic levitation'. *Journal of Electromagnetic Waves and Applications*. 2001;**15**(6):815–831

[10] Kraus, J.D. *Electromagnetics*. Singapore: Mcgraw-Hill Book Co. Ltd.; 1988

[11] Samper, V., Sangster, A.J., Reuben, R.L., & Wallrabe, U. 'Advanced LIGA technology for the integration of an electrostatically controlled bearing in a wobble motor'. *IEEE Journal of Microelectromechanical Systems*. 1998;**7**(4):423–427

[12] Sperry Gyroscope Co. *Microwave transmission design data*, Publication No. 23-80, Sperry Gyroscope Co., 1944; pp. 182–208

[13] Spangenberg, K.R. *Vacuum tubes*. New York, NY: McGraw-Hill; 1948

[14] Eastman, A.V. *Fundamentals of vacuum tubes*. London: McGraw-Hill; 1949

[15] Parker, P. *Electronics*. London: Edward Arnold Ltd.; 1960

[16] Sangster, A.J., & Kelly, M. 'Cylindrical re-entrant cavity resonator design using finite element simulation'. *Microwave & Optical Technology Letters*. 1998;**18**(2):112–117

[17] Terman, F.E. *Electronic and radio engineering*. New York, NY: McGraw-Hill Book Co.; 1955

[18] Samper, V., & Sangster, A.J. 'Torque evaluation of a LIGA fabricated electrostatic micromotor'. *IEEE Journal of Microelectromechanical Systems*. 1999;**8**(1):115–123

Chapter 9
High-frequency magnetic levitation

Engineering is the science of economy, of conserving the energy, kinetic and potential, provided and stored up by nature for the use of man. It is the business of engineering to utilize this energy to the best advantage, so that there may be the least possible waste.

William A. Smith, 1908

9.1 Magnetic levitation in micro-devices

9.1.1 Tuned circuit stabilisation

Magnetic levitation, which is produced, as we have seen in Chapter 3, by the Lorentz forces associated with magnetic induction, has been well known for many years in connection with lifting large masses [1]. On the other hand, the application of magnetic methods to bearings and micro-devices is a relatively recent undertaking [2–5]. The research described in references 2 and 3, for example, was directed towards the creation of a frictionless bearing for a compact gyroscope, and together with [4], these papers are among the earliest to consider 'tuned circuit' stabilisation. In fact, it preceded and informed the tuned circuit developments described in Chapter 8.

A simple sensorless suspension concept based on magnetic forces is illustrated in Figure 9.1. It shows a single degree of freedom arrangement stabilised by means of a series resonant LCR circuit. The levitating electromagnet comprising windings on an E-shaped ferrite core forms the inductor of the circuit. The inductance L is inevitably very large, typically 40–50 mH, and this has considerable impact on the high-frequency operation of the system, as we shall see. Resonance for the LC circuit occurs where (see (8.9)) the inductive reactance equates to the capacitive reactance, or where the inductive stored energy ($W_m = \frac{1}{2}LI^2$ Joules) is balanced by the capacitive stored energy ($W_e = \frac{1}{2}CV^2$ Joules). If C is made very small with the aim of achieving a high operating frequency and compactness of the system, the balancing of W_e and W_m eventually results in a capacitor voltage that exceeds the breakdown level of the dielectric. Typical capacitance values quoted in the literature are quite large – of the order of 2 μF. The implication is that a simple tuned magnetic levitator is forced to operate with a very low f_0 by comparison with the equivalent electric case – in the kHz range rather than MHz.

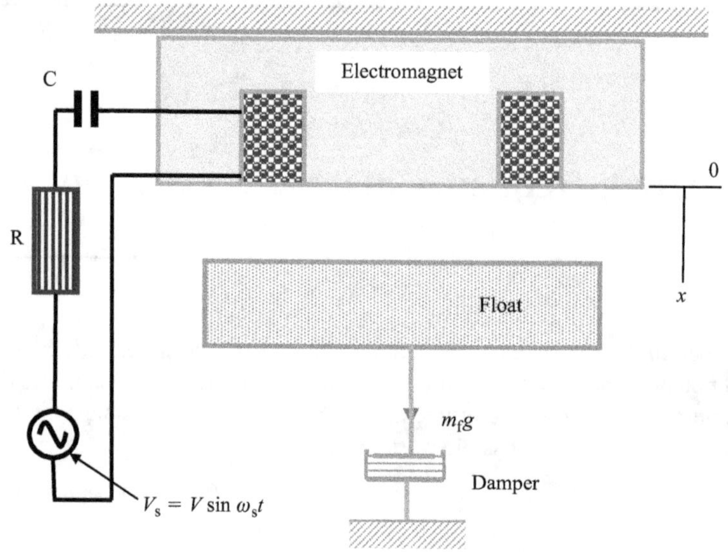

Figure 9.1 Magnetic suspension systems with tune RLC circuit stabilisation

The resonance and tuning characteristics of the series LCR circuit in Figure 9.1 are rather different to those of the electric field levitator described in Chapter 8. This is perhaps not surprising given the large inductive influence exercised by the electromagnet. The float, which is necessarily formed from a ferrous material, is part of the magnetic circuit of the electromagnetic and produces small inductive changes as it moves closer to, or further away from, the primary core. The sensitivity of the resonant frequency to this movement can be established in a manner not too different from the electric field levitator. As in (8.9), the impedance of the series circuit has the form:

$$Z_s = R + j\omega_s L + \frac{1}{j\omega_s C} \tag{9.1}$$

Hence given that the current traversing the circuit i is related to the source voltage V_s through

$$i = \frac{V_s}{Z_s} \tag{9.2}$$

and that the voltage across the terminals of the coil of the electromagnet V_L is given by

$$V_L = j\omega_s L i \tag{9.3}$$

It follows that

$$V_L = \frac{j\omega_s L V_s}{Z_s} \tag{9.4}$$

Hence, with a modicum of algebraic manipulation, it is not too difficult to demonstrate that

$$\left|\frac{V_L}{V_s}\right| = \frac{\omega_s^2 LC}{\sqrt{(1-\omega_s^2 LC)^2 + \omega_s^2 C^2 R^2}} \tag{9.5}$$

Using inductance, capacitance and resistance values from reference 7, which gives

$$\frac{CR^2}{L_0} = 0.0003$$

the voltage ratio in (9.5) is plotted in Figure 9.2 as a function of normalised frequency $\omega_s\sqrt{L_0C}$. L_0 is the inductance of the electromagnet when the float is at its equilibrium position. Note that the vertical scale is logarithmic reflecting the large voltage range between resonance and off-resonance frequencies. It is much larger than in Figure 8.3. The Q of the tuned magnetic levitator is clearly much higher than in the electric field levitator (Chapter 8). This means that the system is very sensitive to float movement, which can lead to instability, because even small changes in position can result in the restoring force falling so far that it enters a negative damping regime [7]. The float begins to oscillate with growing amplitude. This problem is exacerbated by the non-linear properties of the electromagnet. The circuit Q could, of course, be reduced by increasing the resistance (R). But this would mean wastefully using power just to heat the resistance, negating the search for the efficiency that levitation potentially provides.

An alternative form of stabilisation, which has been explored in the literature [5–7], involves the use of a damper attached to the float (as suggested in Figure 9.1)

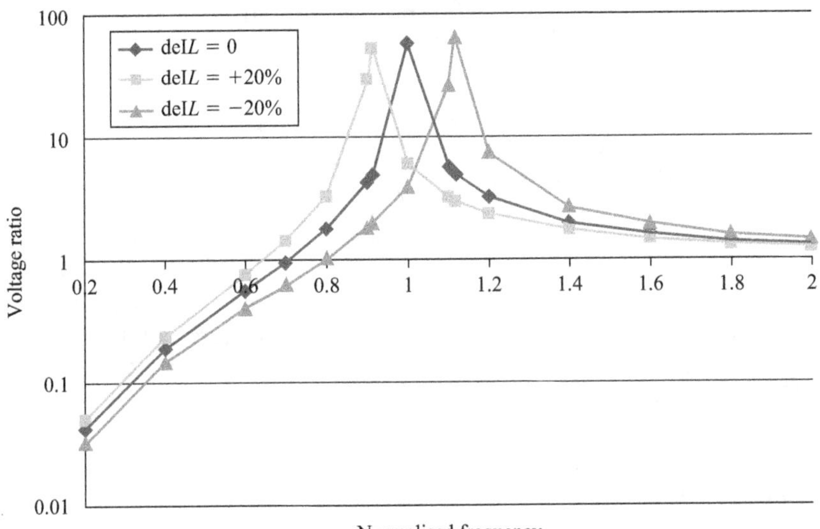

Figure 9.2 Voltage across electromagnet divided by source voltage as a function of normalised frequency $\omega_s\sqrt{L_0C}$

to suppress the divergent vibrations. This is hardly satisfactory since the float is no longer 'free' and the damping mechanism, however realised, becomes a source of drag and friction.

9.1.2 Float dynamics

The levitation forces and the dynamics of the float in the scheme depicted in Figure 8.1 can be deduced as follows. The force on the float is related to changes in stored energy. That is:

$$F_{\mathrm{eq}} = \frac{\mathrm{d}W_{\mathrm{m}}}{\mathrm{d}x} = \frac{\mathrm{d}[\frac{1}{2}L_{\mathrm{eq}}i^2]}{\mathrm{d}x} = -\frac{1}{2}\frac{\mathrm{d}L_{\mathrm{eq}}}{\mathrm{d}x}i^2 \tag{9.6}$$

Here, $\mathrm{d}x$ represents the vertical float disturbance, while L_{eq} and F_{eq} are, respectively, the equilibrium values for the inductance of the electromagnet and the force on the float. i is the instantaneous current in the RLC circuit. Also, using (9.1) and (9.2), the circuit relations can be expressed in the alternative time varying form [7]

$$L_{\mathrm{eq}}\frac{\mathrm{d}^2 i}{\mathrm{d}t^2} + R\frac{\mathrm{d}i}{\mathrm{d}t} + \frac{i}{C} = \frac{\mathrm{d}V_{\mathrm{s}}}{\mathrm{d}t} = \omega_{\mathrm{s}}V\cos\omega_{\mathrm{s}}t \tag{9.7}$$

for a sinusoidal source voltage $V_{\mathrm{s}} = V\sin\omega_{\mathrm{s}}t$. The solution to (9.7) is

$$i = A_{\mathrm{eq}}\sin\omega_{\mathrm{s}}t + B_{\mathrm{eq}}\cos\omega_{\mathrm{s}}t \tag{9.8}$$

where

$$A_{\mathrm{eq}} = \frac{VR}{\rho^2 + R^2} \tag{9.9}$$

$$B_{\mathrm{eq}} = \frac{V\rho}{\rho^2 + R^2} \tag{9.10}$$

$$\rho = \frac{1 - \omega_{\mathrm{s}}^2 L_{\mathrm{eq}}C}{\omega_{\mathrm{s}}C} \tag{9.11}$$

Combining (9.6) and (9.8) yields:

$$F_{\mathrm{eq}} = -\frac{1}{4}\frac{dL_{\mathrm{eq}}}{dx}\frac{V^2}{\rho^2 + R^2} \quad \text{(Newton)} \tag{9.12}$$

At resonance, a typical levitator of this type [7] would produce a force of the order of 3.5 kN. This very large static force at equilibrium is indicative of the impressive levitation potential of even relatively modest magnetic systems (see Chapter 4). At any other condition of the system, (9.7) has to be solved in conjunction with the dynamic force equation, namely

$$m_{\mathrm{f}}\frac{\mathrm{d}^2 x}{\mathrm{d}t^2} = F_{\mathrm{d}} + m_{\mathrm{f}}g - F \tag{9.13}$$

where F_{d} is a disturbance force, F is the instantaneous levitation force and m_{f} is the mass of the float. These equations have been solved in reference 7 and have led to the observation that a tuned LC circuit magnetic suspension system, as depicted in Figure 9.1 is inherently unstable in dynamic terms, unless additional damping is applied to the float.

Given the obvious disadvantages of direct damping of the float, Jin and Higuchi [7] have examined a slightly modified system (Figure 9.3) where the damper is applied between a suspended electromagnet and ground. Hence, on solving a slightly adjusted version of (9.13), together with the circuit equations, they demonstrate that for this modified geometry the tuned LC circuit provides stable levitation. This is illustrated in Figure 9.4, where the modelled response of a typical suspension system, to a 10 N step disturbance to the float, is presented.

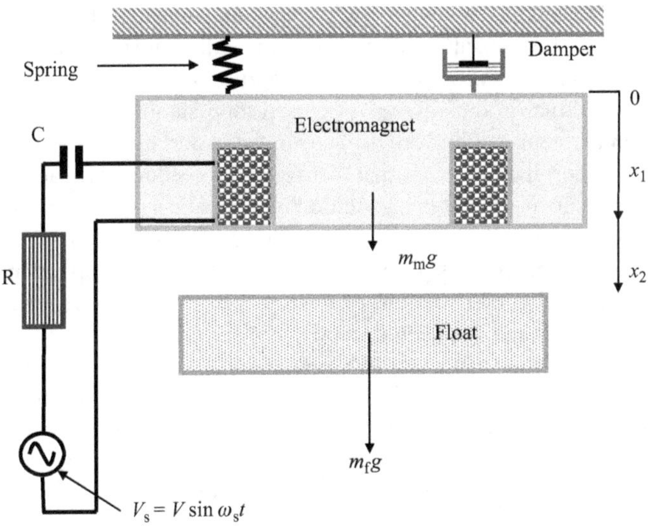

Figure 9.3 Tune LCR magnetic suspension system with damping

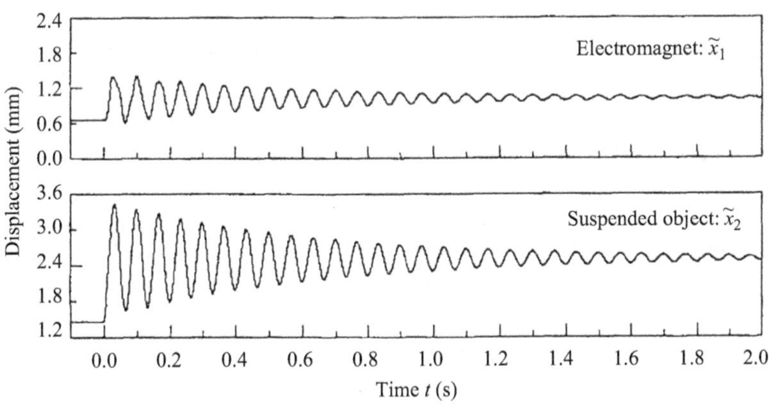

Figure 9.4 Response of damped magnetic levitator to step disturbance

Thus, while stable magnetic levitation of a free float, using a tuned LCR circuit within the electrical supply network, has been shown to be possible, the system components, suggested by Figure 9.3, result in very low operating frequencies and long time constants. Clearly such systems are not conducive to high-frequency operation, which is a prerequisite of miniaturisation and integration into electrical devices in the MEMS category. The down-sizing problem is only exacerbated by the need to employ damping.

9.1.3 Application to micro-devices

As we have already seen in Chapter 8, at the millimetre scale of dimensions high frequencies are needed to generate useable levitation forces. Furthermore, radiation and leakage makes the realisation of tuned circuits with realistic Q-levels difficult to achieve. In addition, even with tuned circuit stabilisation incorporating damping, it has been observed that in the absence of closed-loop stability augmentation, magnetic suspension systems suffer from drift attributed in part to the non-linearity of the solenoid [6]. The corollary to this is that there is still a need for feedback sensing and control electronics to achieve stable levitation in magnetic suspension systems of the above type and this represents a considerable disadvantage to their implementation at the micro-scale of dimensions. While recent research [4], as described in Chapter 4, has provided practical demonstrations of micro-scale magnetic suspension, thus far successful implementation has been dependent on the adoption of a complex closed loop control arrangement to ensure stable operation.

The thrust of this chapter is to provide an assessment of frequency-stabilized magnetic levitation systems that display a potential for miniaturisation. Clearly, tuned-LCR circuit techniques when applied directly to electromagnet levitator are unlikely to permit significant miniaturisation possibilities. As with capacitive systems, described in Chapter 8, the lumped circuit elements resist down-sizing because of wavelength-dependent loss mechanisms. In Chapter 8, the natural evolution of a stable levitator based on a detuned resonator solution, which does not suffer these shrinkage problems, is reported. It is based, not on an LCR circuit, but on an electromagnetic cavity. In the following sections, a similar scheme that has been evolved using magnetic rather than electric forces in a cavity resonator is described.

9.2 Cavity levitation

The detuned magnetic field suspension system [2] described in section 9.1.3 is essentially the analogue of the arrangement described in Chapter 8 [8–10], differing only insofar as the levitation forces are generated within the inductive element of the tuned system rather than the capacitive element. The latter system, which chronologically precedes the magnetic version outlined in section 8.1, has the advantage of circumventing the non-linearities associated with tuned magnetic suspension systems, and this is also true (as we saw in Chapter 8) of the alternative high-frequency detuned levitation systems based on the use of re-entrant cavity resonators. By employing microwave, or millimetre wave, coupled cavity resonators, rather than lumped LC tuned circuits, it is shown in Chapter 8 [11–13] that the same detuned resonator technique can, in principle, be used to produce stable levitation in devices at the micro-scale of dimensions. It is demonstrated there that this cavity resonator technique relies on the use of cavities supporting modes that demonstrate good isolation between the desired electric field suspension forces and the opposing Lorentz forces generated by the magnetic field component of the cavity mode [13]. Patently, the question that arises at this juncture is: Can the cavity technique be reversed to favour Lorentz forces over electric field forces?

Studies on the forces exerted on the walls of both standing-wave and travelling-wave cavity resonators, reveal that for most trapped modes, the Lorentz forces associated with the magnetic field component, are actually more easy to isolate than the electric field forces [13]. Furthermore, as will be shown later in this section, these forces are generally more uniformly distributed over the cavity walls for relatively conventional cavity shapes and modal patterns. This implies that, in principle, cavity levitation based on the Lorentz force exhibits considerable potential as a means of securing miniature stable suspension systems. The chapter provides a detailed examination of a tuned magnetic suspension technique, which harnesses these modal Lorentz forces acting on a 'floating wall' of a metallic electromagnetic cavity resonator. Both disc-shaped and cylindrically shaped 'floats', which form part of the wall of a microwave cavity, are described.

9.2.1 Magnetic force calculations

In Chapter 8 it has been shown, by applying the electromagnetic wave equation to a rectangular waveguide boundary value problem, that the field distribution for the TE_{101} mode, in a waveguide closed at both ends, can be expressed in the form

$$\mathbf{E} = \hat{\mathbf{a}}_y E_0 \sin\frac{\pi x}{a}\sin\frac{\pi z}{L} \tag{9.14}$$

$$\mathbf{H} = \hat{\mathbf{a}}_x \frac{j\bar{\beta}_{10}}{\omega\mu_0} E_0 \sin\frac{\pi x}{a}\cos\frac{\pi z}{L} + \hat{\mathbf{a}}_z \frac{j\pi}{\omega\mu_0 a} E_0 \cos\frac{\pi x}{a}\sin\frac{\pi z}{L} \tag{9.15}$$

where $\hat{\mathbf{a}}_x$, $\hat{\mathbf{a}}_y$ and $\hat{\mathbf{a}}_z$ are unit vectors and β_{10} is the propagation coefficient for the z-directed waveguide mode that forms the cavity resonance. The field pattern is depicted in Figure 9.5. Close examination of the above field equations, or alternatively the sketched pattern in Figure 9.5, shows that on the side walls at $x = 0$ and $x = a$, $E_y = 0$ while H_z is finite, or on the end walls at $z = 0$ or $z = L$ (see Chapter 2), $E_y = 0$ and H_x is finite. On these walls, therefore, the Lorentz force is isolated from the electric field force that exists only on the top and bottom walls at $y = 0$ and $y = b$. This force can be determined by examining the change in magnetic stored energy that occurs when the wall of interest is perturbed by an infinitesimal amount. It is assumed in so doing that the modal fields remain essentially undisturbed.

For the end wall at $z = L$ (say), the H-field at this wall is

$$H_x = \frac{j\bar{\beta}_{10}}{\omega\mu_0} E_0 \sin\frac{\pi x}{a} \tag{9.16}$$

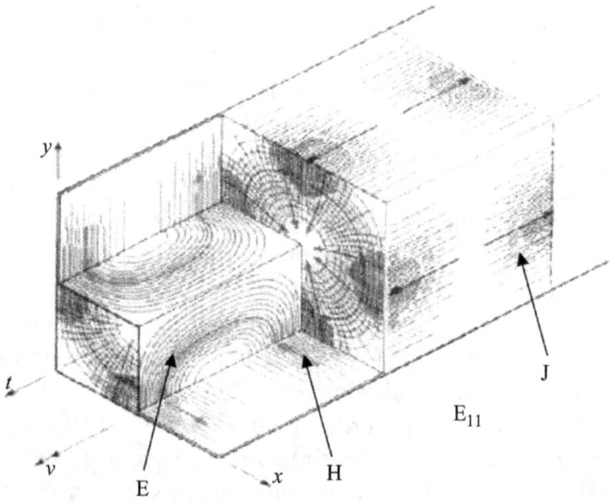

Figure 9.5 *TM_{11} mode in rectangular waveguide – E, H and J patterns (see personal.ee.surrey.ac.uk/Personal/D.Jefferies/wguide.html)*

Now, if F_m is the average time Lorentz force is acting on this wall, then the work done in moving it a distance dz in z-direction is $W_m = F_m\,\mathrm{d}z$. But W_m must equal the change in energy in the cavity caused by this adjustment. Given that the time averaged energy density in the cavity for magnetic field (peak value) is given by

$$U_{\mathrm{m}} = \frac{\mu_0 |H^2|}{4} \quad (\mathrm{J/m^3}) \tag{9.17}$$

The time average work done on the movable end wall is then given by

$$F_{\mathrm{m}}\,\mathrm{d}z = \int_0^b \int_0^a \frac{\mu_0 |H_x^2|}{4} \mathrm{d}x\,\mathrm{d}y\,\mathrm{d}z \quad (\text{Joules}) \tag{9.18}$$

And hence the repulsive force on the wall is

$$F_{\mathrm{m}} = \frac{\bar{\beta}_{10}^2 ab}{8\omega^2 \mu_0} E_0^2 \quad (\text{Newtons}) \tag{9.19}$$

The magnitude of this force in a rectangular cavity with $a = 2$ cm, $b = 1$ cm and $\omega = 62.83 \times 10^9$ rad/s can be estimated if the cavity Q and absorbed power are known. In a cavity with high conductivity walls, Q could be the order of 10,000 and a power absorbed $P = 1$ W is not untypical. Given that, in general, the energy stored in a cavity (W_{cav}) is related to P and Q through the relation

$$W_{\mathrm{cav}} = \frac{PQ}{\omega} \quad (\text{Joules}) \tag{9.20}$$

it is not difficult to compute that in such a cavity the peak electric field will be $E_0 = 0.355$ kV/cm. Hence from (9.19), $F_m = 1.2$ nN, which may seem extremely small but it is not insignificant in micro-actuator terms.

Unfortunately, if the end wall of the above cavity is not in good electrical contact with the other walls, as it would be if flotation was intended, then even this tiny force would not be available. The narrow gaps between an end wall float and the main cavity block becomes a strong source of radiation. The cavity becomes very lossy and the high Q resonance is quenched.

Successful levitation, using the Lorentz force produced by the fields of a cavity resonator, demands that the free wall can be introduced without disturbing the modal fields and hence the resonant behaviour of the cavity. This essentially requires that the air gap between the free wall, or float, runs parallel to the current density flows of the target mode.

9.2.2 TM_{11} mode ring resonator

The TM_{11} mode is an example of a rectangular waveguide mode, which continues to propagate unchanged or undiminished by axial gaps disconnecting the side walls from the top and bottom walls. The field patterns and current density flows for this mode are depicted in Figure 9.5. The directional distributions of the electric field and the magnetic field in the guide interior and the current density in the guide walls, are identified by pointers on the diagram. Note that the loops of magnetic field that lie wholly on the x–y plane at all cross-sections, forces the surface current density, which must be orthogonal to the magnetic field at the conductor surface, to be z-directed everywhere. Consequently, a narrow axial (z-directed) slit in any wall will have little effect on the mode since the current flows are undisturbed. The corollary is that freeing one of the walls from its neighbours will not quench the mode provided the gap can be considered to be a narrow slit.

Electromagnetic field studies on the TM_{11} mode as a possible source of levitation in micro-scale devices have mainly been directed [13] at the rectangular waveguide ring resonator. The reasons behind this choice are manifold. First, when operated in a TM mode (TM_{11} in this case), the waveguide supports circumferential wall currents, as identified in Figure 9.5, which means that the TM mode is not significantly disturbed by loss of contact between the floating wall forming the top wall of the ring and the resonator base. A dimensioned drawing of a device designed to operate at 11.5 GHz is presented in Figure 9.6. The ring-shaped 2.98 cm wide rectangular waveguide channel is machined into a solid aluminium disc to a depth of 1.49 cm. The second reason for choosing the rectangular cross-section ring resonator is that it is relatively easy to couple to the TM mode through a transverse (radial) slot in the waveguide. The single slot coupler is shown on the left of Figure 9.6, with the feed waveguide channel shown below it. More than one slot is possible to procure directional coupling. Third, the travelling-wave ring mode potentially provides uniform levitation forces around the ring as we shall see.

9.2.3 Ring resonator levitation force

The levitation mechanism can be illustrated, and the force available as a function of power and frequency can be determined, by first considering the magnetic field expressions for a TM_{11} mode in a rectangular waveguide with major dimension a and

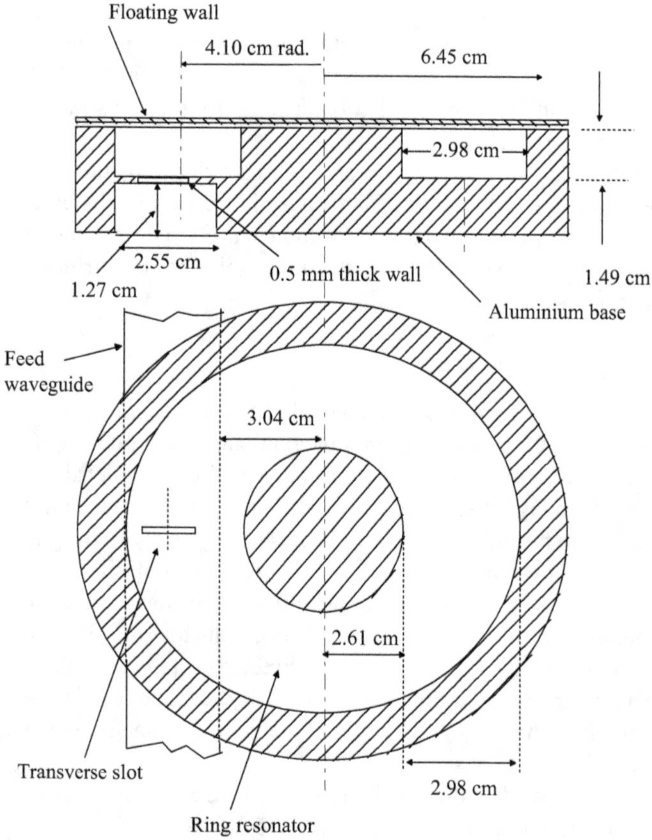

Figure 9.6 Schematic illustration of a ring resonator used in levitation experiments

minor dimension b. This can be done by following the procedures in Chapter 2 for a TM rather than a TE mode, and then seeking the TM_{11} solution. The process leads to

$$\begin{aligned} H_x &= \frac{\omega\varepsilon}{\beta_{11}} E_0 \sin\frac{\pi x}{a} \cos\frac{\pi y}{b} \exp(-j\beta_{11}z) \\ H_y &= -\frac{b}{a}\frac{\omega\varepsilon}{\beta_{11}} E_0 \cos\frac{\pi x}{a} \sin\frac{\pi y}{b} \exp(-j\beta_{11}z) \end{aligned} \tag{9.21}$$

where $\beta_{11}^2 = k^2 - \frac{\pi^2}{a^2} - \frac{\pi^2}{b^2}$, and x, y and z are directed along the a, b and axial waveguide directions (see Figure 2.5). E-field magnitude patterns for the TM_{11} mode in a rectangular waveguide, formed into a ring resonator, are shown in Figure 9.7. These were generated using Ansoft's high-frequency structure simulator [14]. The TM_{11} mode moulded into a ring geometry has the advantage of naturally fitting into potential bearing and micromotor configurations, but is by no means the only possibility. The pattern depicted in Figure 9.7a is of the TM_{111} mode with the mean distance around the ring being one wavelength, while in Figure 9.7b the TM_{11} mode

(a)

(b)

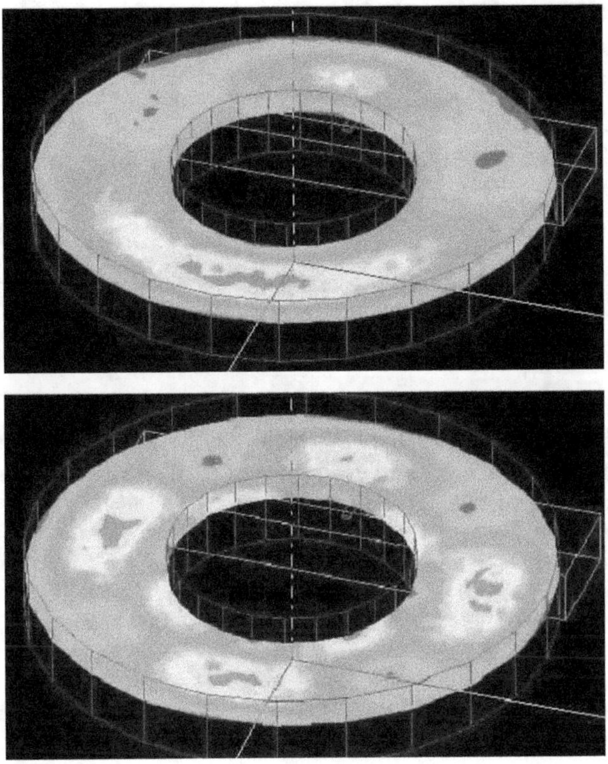

Figure 9.7 Simulated TM modes in a ring resonator: (a) TM_{111} mode (b) TM_{112} mode

is 'ringing' at exactly two wavelength forming the TM_{112} mode. In the field patterns dark irregular patches denote high field strength, while dark circular spots represent regions of very low E-field. The slight asymmetry, particularly in Figure 9.7a, is due to the presence of the excitation slot.

The transverse magnetic fields, formulated in (9.21), generate purely axial currents in the walls of the waveguide, as required by the boundary conditions at a metallic interface. In the presence of these currents, the modal magnetic fields exert a 'repulsive' Lorentz force on the walls. For an elemental distance dz in the axial direction within the waveguide, the elemental force is

$$\mathrm{d}F = Bi\,\mathrm{d}z \tag{9.22}$$

where B represents the transverse magnetic flux density at a wall, and i denotes the current in the wall at the same position. At the floating top wall of the waveguide (Figure 9.6) we can write

$$B = \mu_0 H_x(x, z) \tag{9.23}$$

$$i = J_z(x, z)\mathrm{d}x = H_x(x, z)\mathrm{d}x \tag{9.24}$$

Combining (9.22), (9.23) and (9.24), and time averaging, gives

$$dF_z = \frac{1}{4}\mu_0 H_x^2(x,z)dA \tag{9.25}$$

where $dA = dx\,dz$ is an element of area in the top wall.

Over a length L of the floating wall, the full Lorentz force is thus

$$F_m = \frac{1}{4}\int_0^L \int_0^a \mu_0 H_x^2(x,z)\,dx\,dz \tag{9.26}$$

(cf. (9.18)). Using (9.21), and on processing the integrals, (9.26) thus becomes

$$F_m = \frac{\omega^2 \mu_0 \varepsilon^2 E_0^2}{16\beta_{11}^2} aL \quad \text{(Newtons)} \tag{9.27}$$

The TM_{11} mode electric field is normal to the top and bottom walls of the cavity with a distribution analogous to the TE_{10} mode. This field produces an attractive force F_e that has already been evaluated in Chapter 8 (8.16). We can use this equation for the TM_{11} mode if we set $w = a$ and $l = L$. Hence

$$F_e = \frac{\varepsilon_0 E_0^2}{16} aL \quad \text{(Newtons)} \tag{9.28}$$

The ratio of these two forces, which is critical to levitation becomes

$$\frac{F_m}{F_e} = \frac{\omega^2 \mu_0 \varepsilon_0}{\beta_{11}^2} = \frac{k_0^2}{\beta_{11}^2} = \left[1 - \frac{f_{c11}^2}{f^2}\right]^{-1} \tag{9.29}$$

This simple relationship is plotted in Figure 9.8 as a function of f/f_{c11}. In the assumed that for 2 cm × 1 cm waveguide, the cutoff frequency is 16.77 GHz. It is clear from this figure that the Lorentz force predominates very strongly over the electric force for frequencies close to cut-off. In ring resonator terms, this implies that a

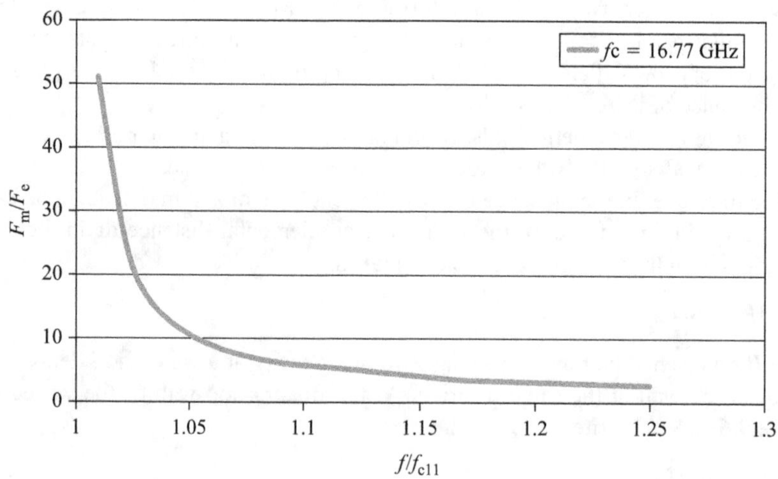

Figure 9.8 Ratio of Lorentz to electric field force in TM_{11} mode ring cavity

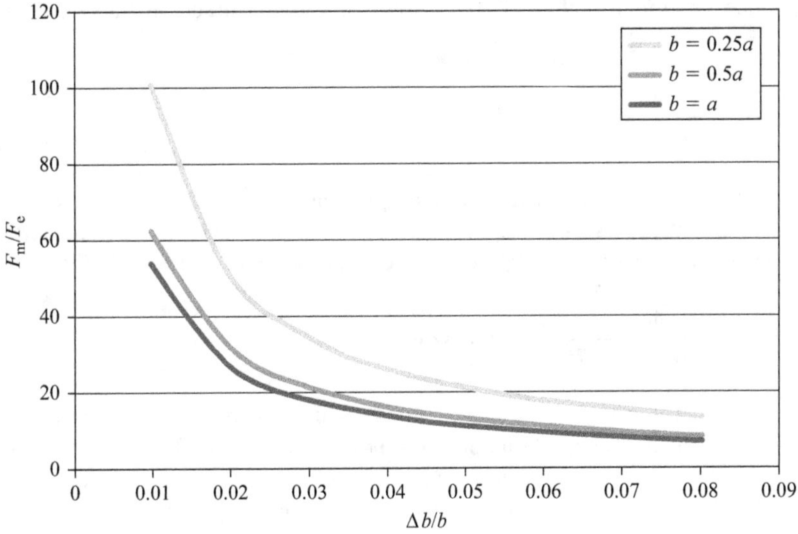

Figure 9.9 Lorentz to electric field force ratio as a function of guide height b for TM_{110} mode in ring waveguide with a = 2 cm

levitator design should aim to employ the lowest possible modes such as the TM_{110} or TM_{111}. Higher-order modes would dictate the use of rings of increasingly large radii to ensure that the desired mode remains close to cut-off, and hence retain a high value of F_m/F_e.

The sensitivity of the Lorentz force relative to the opposing electric field force to the movement of the free wall, or float, is presented in Figure 9.9, for the TM_{110} mode. It is assumed that the float is at $y = b$ (see Figure 9.5) and the movement Δb is expressed as a ratio with b. Three cases are examined, $b = 0.25a$, $b = 0.5a$, $b = a$, for a guide with $a = 2$ cm. The trend for all three curves is rather similar to that shown in Figure 9.8, which is not too surprising since increasing b lowers the cut-off frequency of the guide. The movements suggested by the horizontal axis range, namely a few percent of the guide height before the Lorentz force is seriously diminished, is not untypical of likely float movements in stabilised micro-scale levitation devices.

9.2.4 Power requirements

For a TM_{11} mode cavity of length L, the energy stored in the cavity, as we have already seen, is

$$\begin{aligned} W_m &= \frac{1}{4}\int_0^a\int_0^b\int_0^L \mu_0|H|^2\,dV \\ &= \frac{\omega^2\mu_0\varepsilon^2}{4\beta_{11}^2}E_0^2\left[\frac{b^2}{a^2}+1\right]\frac{abL}{8} \quad \text{(Joules)} \end{aligned} \tag{9.30}$$

Hence, assuming that the Lorentz force is strongly predominant (e.g. TM_{110} mode)

$$F_{\text{lev}} = \frac{4a^2}{b(a^2+b^2)} W_{\text{m}} \tag{9.31}$$

But for any resonator, stored energy $W_{\text{m}} = \frac{PQ}{2\pi f_0}$, where P is the power delivered to the resonator in Watts, Q is its quality factor and f_0 is the resonant frequency in Hertz.

For the TM_{110} mode $f_0 = f_{c11}$, and F_{lev} can therefore be expressed in the form:

$$F_{\text{lev}} = \frac{4a^2}{b(a^2+b^2)} \frac{PQ}{2\pi f_{c11}} \quad \text{(Newtons)} \tag{9.32}$$

The levitation force is plotted against power delivered to the cavity resonator in Figure 9.10, for a resonator supporting a TM_{110} mode at 11.3 GHz and displaying a typical Q of 10,000. The curve is plotted for cavities of three different heights ($b = 0.5$ cm, 1.0 cm and 1.49 cm). The figure shows that levitation forces in the 0.001–1.0 mN range are readily available with modest drive powers of between 10 mW and 10 W. Note that 1 mN is approximately the force required to lift a 0.1 g weight under the influence of gravity.

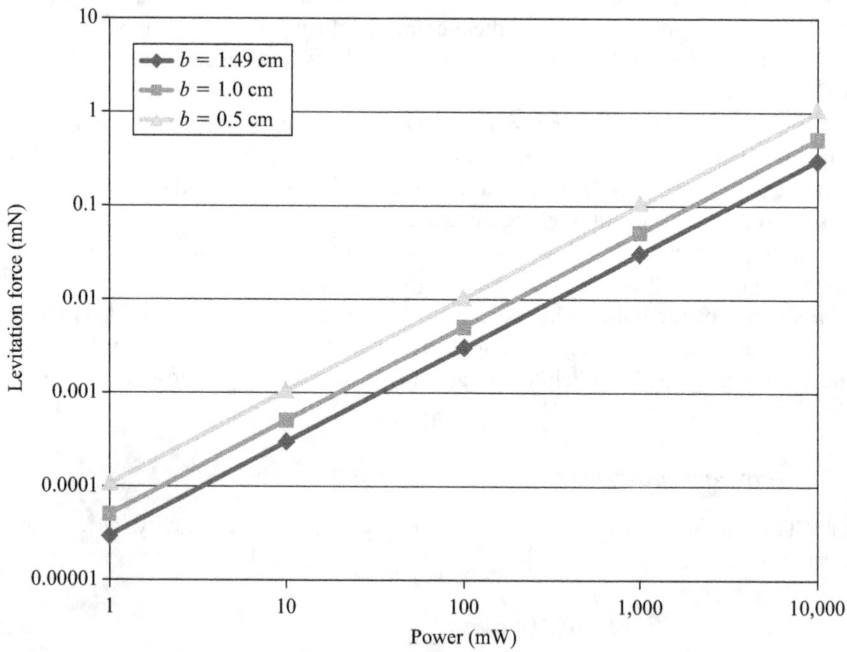

Figure 9.10 Estimated levitation force versus power delivered for a TM_{110} mode ring resonator with $a = 2.98$ cm, $Q = 10{,}000$ and $f_{c11} = 11.3$ GHz

For a TM_{110} mode ring resonator, the ratio of the available levitation force to the gravitational force on the floating wall can readily be estimated. The gravitational force is given by

$$F_{grav} = \rho a L t g \quad \text{(Newtons)} \tag{9.33}$$

where ρ is the density of the 'float', t is its thickness and g is gravitational acceleration. When (9.32) and (9.33) are combined and using the following dimensional approximations, $a \approx \frac{\lambda_{c11}}{2}$, $b \approx \frac{\lambda_{c11}}{4}$ and $L \approx 4\lambda_{c11}$, the following relationship between the levitation and gravitational forces is obtained:

$$\frac{F_{lev}}{F_{grav}} \approx \frac{16 f_{c11}^2 PQ}{5\pi \rho g c^3 t} \tag{9.34}$$

This force ratio is presented as a function of plate thickness t (microns) in Figure 9.11 for three different power levels P in mW. It shows clearly the advantage of operating at frequencies in the microwave and mm-wave ranges. In the microwave range (f_{c11} = 11.3 GHz), a force ratio in excess of 10 can be achieved for a 0.1 mm thick aluminium 'float' with drive powers of little more than 1 W, assuming the cavity exhibits a Q of at least 10,000. Clearly, with an f_{c11} squared multiplier in the numerator of (9.34), the levitation of a free-wall float on a TM_{11} mode cavity resonator can be secured at much lower power levels than this.

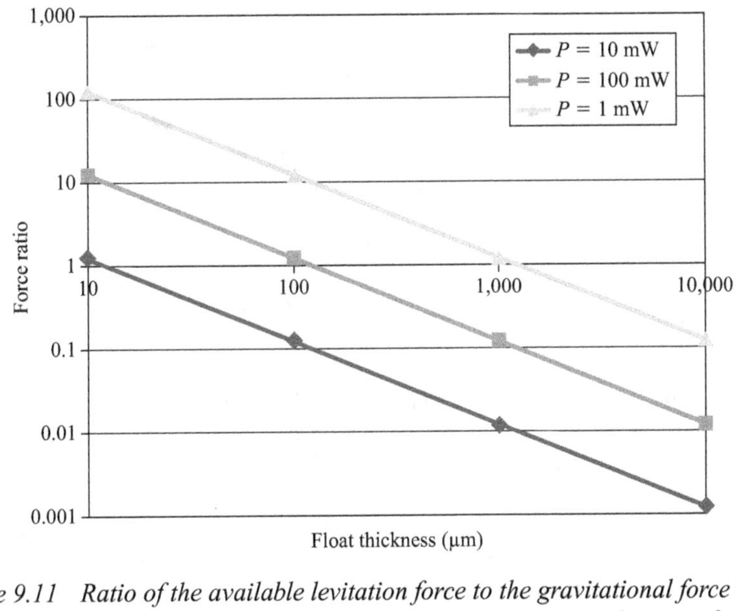

Figure 9.11 Ratio of the available levitation force to the gravitational force versus thickness of aluminium floating wall in a TM_{110} mode cavity for three source power levels (Q = 10,000, $\rho \approx 4.0$ g/cm^3)

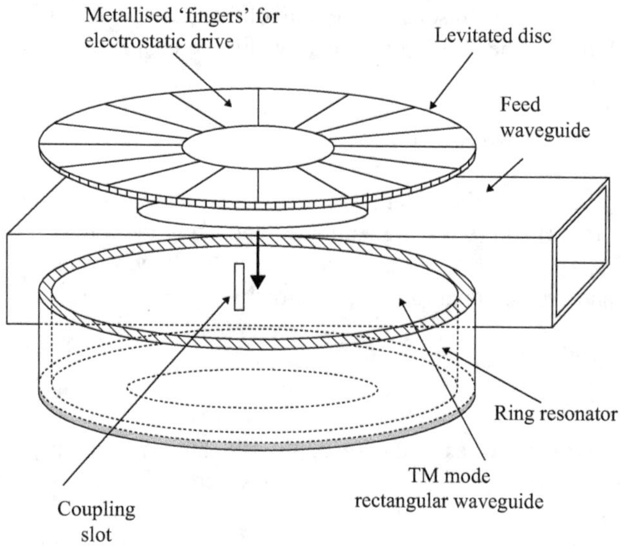

Figure 9.12 Electrostatic motor armature levitated by TM mode ring resonator

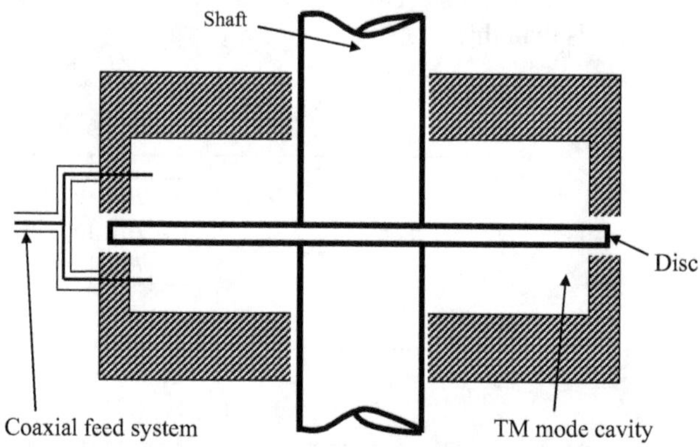

Figure 9.13 Frictionless bearing using TM mode ring resonator

Concept drawings of possible TM ring mode levitators are shown in Figures 9.12 and 9.13. The electrostatic motor bearing depicted in Figure 9.12, if operated at an established mm-wave frequency band in the vicinity of 110 GHz, would employ waveguide with $a = 2.5$ mm and $b = 1$ mm, and support an electrostatic drive disc (float) of about 12 mm diameter. For a device fabricated on silicon with other electronic components, it is unlikely that the mm-wave feed line would be waveguide as suggested. It is much more likely to be microstrip line, or coplanar line, linked to an on-board solid state source. A frictionless bearing for a shaft of a rotating device provided by a TM mode ring cavity is depicted in Figure 9.13. In this case, the levitator need not be particularly small. The shafts from motors varying in size from tiny medical devices to units large enough for kitchen appliances could be envisaged. The primary enabling criterion is likely to be available microwave power, or mm-wave power, at acceptable cost.

9.3 Summary

The tuned magnetic suspension system examined in section 9.1, in addition to being low on stiffness, is difficult to resonate with practical capacitors, because of the high inductance of the electromagnet plus float. Introducing laminations into the magnet and the float to circumvent this problem is highly inconvenient. So while the detuned resonator concept seems attractive, in so far as it provides a seemingly stable force versus distance characteristic between the float and the electromagnet, and hence offers inherent advantages over other systems, rather disappointingly the drawbacks have outweighed these advantages resulting in very little uptake [15].

By employing basic field relations for the dominant TM_{11} mode of a waveguide ring resonator together with fundamental energy expressions for cavity resonators, the primary requirements that must be met by a resonant cavity-based magnetic field levitation system are established. When these requirements are fulfilled stable levitation of a 'free' conducting wall, or 'float' of a waveguide-based ring-resonator supporting a TM resonance is predicted. In particular, it is shown that the available levitating force (Lorentz force) is substantially larger than the gravitational force for TM_{11n} cavity resonators where n is small – ideally zero. For systems designed to operate at mm-wave frequencies (>100 GHz), strong levitation is available even with modest drive powers in the milliwatt range. In fact, experimentation [13] on a TM_{111} mode waveguide ring, resonating in the microwave, rather than the mm-wave range of frequencies, has suggested that a 1 mm thick floating wall weighing 15.88 mg, can be levitated with drive power levels in the 5–15 W range. The TM_{111} mode used in these experiments is reported to have displayed a Q level in excess of 20,000 with the floating wall in its levitated position.

The literature currently paints a picture of a developing technology that has stalled, because applications have failed to materialise, or are being satisfied by other means. However, the current lack of demand for frictionless bearings employing frequency stabilising techniques, particularly for micro-actuators and miniature motors, will surely diminish as efficiency becomes a major goal of technology in an energy-limited world. In small devices, this will also dictate that reliability and long levity will become increasingly desirable properties, and since friction in bearings and moving parts grows inversely with shrinking size, representing a major source of failure in electrical components, the desire for non-contacting moving parts will become very strong. At that point, current impediments to micro-levitator development will surely disappear.

9.4 References

[1] Laithwaite, E.R. *Propulsion without wheels*. New York, NY: Hart Publishing Co. Inc.; 1966, pp. 179–199

[2] Kaplan, B.Z. 'Analysis of a method for magnetic levitation'. *Proceedings of the IEE*. 1967;**114**:1801–1804

[3] Gillinson Jr, P.J., Denhard, W.G., & Frazier, R.H. *A magnetic support for floated inertial instruments*. S.M. Fairchild Publication Fund Paper No. FF-27. Institute of Aeronautical Sciences, New York; May 1960

[4] Williams, C.B., Shearwood, C., Mellor, P.H., & Yates, R.B. 'Modelling and testing of frictionless levitated micromotor'. *Sensors and Actuators*. 1997;**A 61**:469–473

[5] Jayawant, B.V., & Kaplan, B.Z. 'The development and stabilisation of a new electromagnetic suspension'. *Journal of Applied Physics E*. 1971;**4**:301–306

[6] Kaplan, B.Z., & Regev, D. 'Dynamic stabilization of tuned circuit levitators'. *IEEE Transactions*. 1976;**MAG-12**(5):556–559

[7] Jin, J., & Higuchi, T. 'Modelling and dynamic stabilization for suspension system using tuned LC circuit'. *IEEE International Conference on Systems Engineering*; Kobe, Japan, 1992, pp. 411–415

[8] Kumar, S., Cho, D., & Carr, W. 'A proposal for electrically levitating micromotors'. *Sensors and Actuators*. 1990;**24**:141–149

[9] Kumar, S., Cho, D., & Carr, W. 'Experimental study of electric suspension for micro–bearings'. *Journal of MEMS*. March 1992;**1**(1):23–30

[10] He, G.H., Chen, K.S., Tan, S.S., & Wang, W.Y. 'Electrical levitation for micromotors and microaccelerometers'. *8th International Conference on Sensors and Actuators*; Stockholm, Sweden, June 25–29, 1995

[11] Sangster, A.J., & Kelly, M. 'Electromagnetic levitation using the fringing fields of a microwave cavity'. *Proceedings of ANTEM '96*; Montreal, Canada, August 1996, pp. 99–102

[12] Sangster, A.J., McErlean, E., Beale, G., Kelly, M., & Smith, P. 'Coupled re-entrant cavity system for electromagnetic levitation'. *Journal of Electromagnetic Waves and Application*. 2001;**15**(6):815–831

[13] Sangster, A.J., & McErlean, E. 'Electromagnetic levitation of micro-scale actuators using microwave cavity resonators'. *IEE Proceedings of MAP*. 2000;**147**(5):359–363

[14] *Ansoft Corporation HFSS*. Available from http://www.ansoft.com/products/hf/hfss

[15] Jayawant, B.V., & Rea, D.P. 'New electromagnetic suspension and its stabilisation'. *IEE Proceedings*. 1968;**115**(4):549–554

Chapter 10
Sustainability through engineering efficiency

Scientists investigate that which already is; engineers create that which has never been.

Albert Einstein

10.1 Carbonising the atmosphere

Perceptive readers of this volume will have noted that its title is embellished by the sub-heading 'A Key Technology for a New Age of Electricity'. While the author is confident that a textbook purely focused on levitation technology will have a valuable role in the education of coming generations of professional electrical engineers, he is also aware that an impending transition of the global economy, away from fossil fuels towards renewable sources of energy is imminent. This obviously creates an added context to the topics of electricity, and engineering efficiency, which needs to be acknowledged. This chapter is directed at identifying this expanded role and the demands that are likely to be faced by electrical engineers as humankind initiates a 'flight from fossil fuels', and as a consequence, sets in train the creation of an electrically powered future based wholly on renewables. This is a very big issue with complex political and social ramifications [1–11], but we can limit the analysis here to a single chapter by focusing on the efficiency implications of the transition that impact mainly on electrical engineering activities, as we shall see.

At this moment in time, namely Spring 2012, 'business-as-usual' (BAU) for the global economy is the strongly entrenched mindset of political and financial leaders around the world. If this attitude were to persist for another 20 years, and population numbers were to continue to rise towards a plateau predicted to be 10.5 billion by the end of the century, then fossil fuel consumption trends [1] suggest that we will be programmed to release an incredible 25–30 TW of power from coal, oil and gas in 2050. This date is well within the lifetimes of most people on the planet. The trend is shown in Figure 10.1, where the upper curve with diamond markers depicts estimated power consumption for the currently entrenched, energy profligate, BAU scenario. This rate of rise presumes that a slightly diminished rate of growth could occur due to 'peak oil' causing rising energy costs as a consequence of the diminishing liquid fossil fuel reserves after 2020. In energy terms, 25 TW for a year equates to 788×10^{18} J, or 747×10^{15} BTUs. Furthermore, 747×10^{15} BTUs extracted almost wholly from fossil fuels result in approximately 54 Gigatons per year (Gt/year) of carbon dioxide, or 16 GtC/year of carbon, being dumped into the atmosphere by 2050. This is a mean figure, which assumes that the carbon content of coal, oil and natural gas differs in the ratio 1:0.8:0.55. The relative amounts combusted per year also vary significantly. Because of the linearity between power consumed, energy (product of power and time), carbon emitted in GtC and carbon dioxide (3.67 times carbon) in $GtCO_2$, these forms can be used interchangeably, and often are in the scientific literature. For the power trend shown in Figure 10.1, the equivalent rate of increase of atmospheric carbon in GtC/year is shown by the curve with square markers. Of the power consumption displayed in Figure 10.1, half will be expended by industry, a quarter by transport, a sixth by domestic users and a twelfth by commerce [1].

The trend is, of course, unsustainable simply because by continuing on this path, the accumulated CO_2 in the atmosphere becomes extremely dangerous. Cumulated emissions of CO_2 since the industrial revolution have already yielded, as of early 2012, a concentration level in the atmosphere of 393.7 parts per million by volume (ppmv), which is 41% above the long-term background or natural level of 280 ppmv. If the growth pattern of Figure 10.1 is maintained until 2050, the concentration will rise by then to 445 ppmv of CO_2 (ignoring feedbacks) and this, according to the best available science [8,12], equates to an average global temperature rise from pre-industrial levels of close to 3°C. Such a temperature rise in such a short period of time is unprecedented in the paleo-climate record [8] and is predicted to bring about

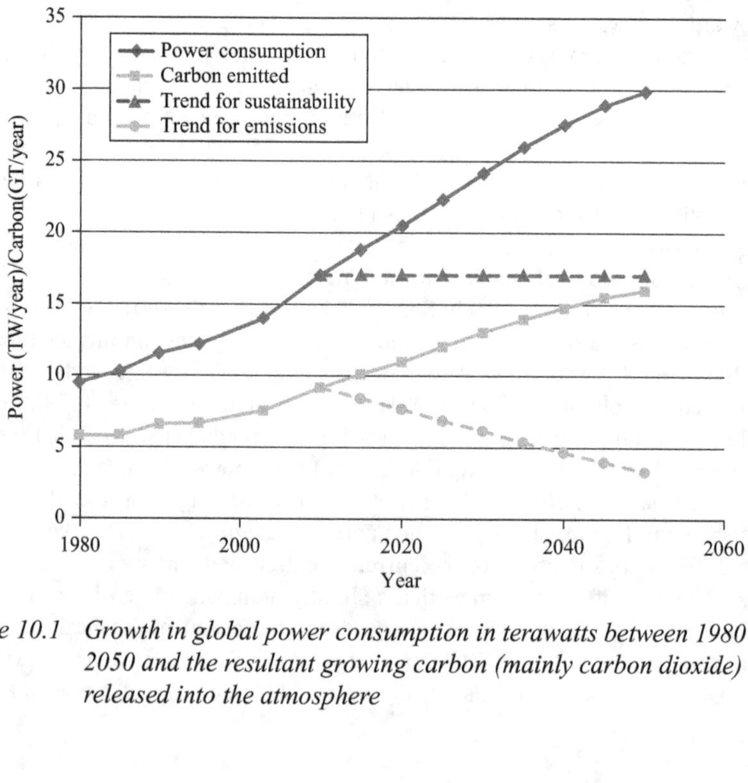

Figure 10.1 Growth in global power consumption in terawatts between 1980 and 2050 and the resultant growing carbon (mainly carbon dioxide) released into the atmosphere

major changes in the planetary climate presaging extremely wild and unpredictable weather around the globe with major disruptions to food production. Evidently, the warming trend has to be halted and reversed. However, few non-scientists are, as yet, prepared to accept that this is necessary, because the implications for the global market, capitalism, consumerism and 'western' lifestyles remain unthinkable.

A rapid transition from fossil fuels to renewable resources, as the primary energy source propelling forward our modern economies, is essential and unavoidable. How this can be engineered will be pursued in ensuing sections.

10.2 Wedging towards decarbonisation

Clearly, the only truly safe concentration of CO_2 in the atmosphere is the level that has prevailed for thousands of years since the last ice age, namely 280 ppmv. This measured result has been gleaned from the chemistry of ice cores that have been carefully extracted from ice sheets in Antarctica and other glaciers around the globe [10]. Realistically however, modern energy hungry capitalist economies are never going to be persuaded to tolerate the degree of recessionary behaviour that will be necessary to return CO_2 concentrations to these pre-industrial levels. Nevertheless, it is certainly possible for advanced human societies to reduce carbon emissions to a rate that slows and reverses the accumulations. The scientific consensus generally hovers around 450 ppmv as the upper boundary to CO_2 pollution of the atmosphere. According to sophisticated computer models of the biosphere, this ensures that the average global temperature increase from pre-industrial levels stays below a 'safe' +2°C. To achieve 450 ppmv, the latest science suggests that emissions of carbon to the atmosphere must drop back to ~3 GtC/year by 2050 (dashed curve with circular markers in Figure 10.1). The emissions, associated with the 'wedge' representing the difference between the BAU curve (solid line square markers) and the dashed line representing the desired emissions trend, have therefore to be addressed and this can only be done by implementing changes in how energy is used. In Figure 10.1, it is postulated that one way to achieve the desired savings would be to flat line (dashed curve with triangular markers), at the 2012 level, the power consumed by the global economy. This does not seem to be too onerous for the world community, but it does call for a rapid cut back in fossil fuel use to procure the emissions trend dictated by the lower curve. The loss of energy from coal, oil and gas will have to be replaced by power from renewables to maintain a flat power generation profile.

The route to procuring the required change, in the energy-use profile of the world economy, has been powerfully addressed by Stephen Pacala and Robert Socolow in an article published in the journal *Science* [13] in 2004. In it they argue that atmospheric CO_2 levels can be stabilised over the long term by implementing a coherent and interwoven range of "stabilization wedges" representing technology adjustment over the next 50 years. The concept is actually quite simple requiring the carbon emissions wedge in Figure 10.1 to be triangulated and divided into a number of smaller triangles: each sub-triangle or wedge represents one in-hand technology or societal practice that could be introduced immediately, relatively slowly at first and increasing

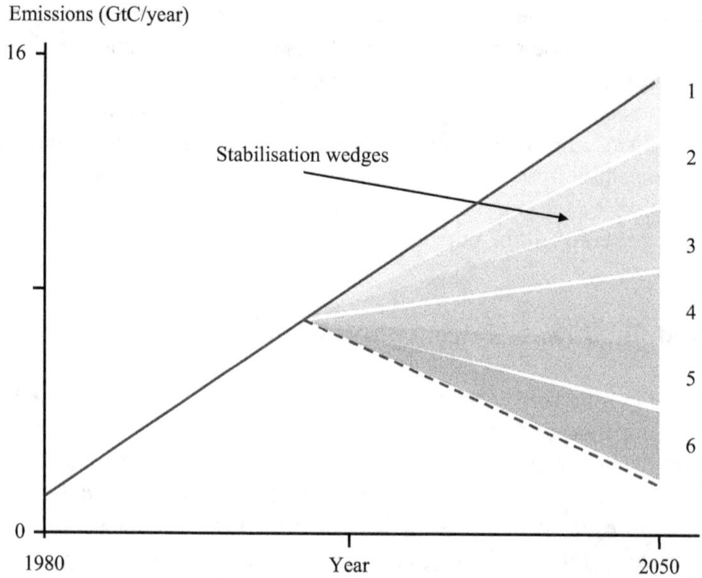

Figure 10.2 Stabilisation wedges as pioneered by Pacala and Socolow [13]

linearly with time, to lower power use from fossil fuels and make a steady decline in the rise in CO_2 emissions. This is depicted schematically in Figure 10.2.

Six wedges are depicted in Figure 10.2: there could be more. Pacala and Socolow initially used seven but increased it to nine 7 years later. They listed out possible 15 interchangeable schemes. Each wedge represent technologies or activities selected from this list of possibilities that include recognised mitigation strategies such as transitioning away from private cars and aircraft towards electrified rail systems, increasing energy efficiencies in vehicles and buildings, nuclear power expansion, reducing deforestation and increasing reforestation and improving agricultural practices. These adjustments in parallel with a concerted and determined drive to expand renewable resources on a worldwide scale could permit global economies to operate at tolerable 2010 power consumption levels while pushing emissions towards zero by mid-century. It is imperative for the health of the biosphere that this emissions drop is achieved in the next 40 years and this, in turn, requires an early commencement on the process of diverting manpower and facilities towards the development and implementation of these technologies. It is the author's belief, that the advances in electromagnetic levitation and superconduction described in this book will feature strongly within the transport and efficiency wedges of an actively transitioning world.

It is perhaps worth noting here that the power stabilisation strategy suggested in Figures 10.1 and 10.2 differs from that of Pacala and Socolow, who sought stabilisation of emissions at 2004 levels, a somewhat less 'painful' readjustment for the global capitalist system, by permitting a risky doubling of the carbon dioxide concentration in the atmosphere. However, little progress has occurred in the intervening years to ameliorate emissions. Consequently, it is now not possible, even if we were to commence now with the Pacala/Socolow strategy adjusted to the aim of maintaining carbon emissions at the 2012 level, thereby procuring their 'lax' goal of limiting CO_2 growth to a doubling from pre-industrial levels (representing a +3°C temperature anomaly). Needless to say, minimising the risk of the global average temperature breaking through the 'safe' +2°C barrier now requires much more strenuous actions to curb emissions.

10.3 Infilling the wedges

The labour market for graduate electrical engineers over the next few years will veer sharply towards the renewables industry and the 'greening' of industrial economies as humanity begins to realise that fossil fuels are running out, but more importantly, that burning fossil fuels is dangerous because of climate change. So it seems appropriate here to identify, as far as is possible, those shifts in societal behaviour that will impinge on the practice of electrical engineering, and as a consequence increase the role of electromagnetically levitated devices and systems. Given that 'a picture is worth a thousand words', some simplistic yet attention seeking illustrations, rather reminiscent of those found in managerial reports to a technologically illiterate board of lawyers, accountants and bankers of a British company, have been incorporated into the text to clearly identify and emphasise the transitions required.

To make serious progress towards the primary aim of reducing carbon emissions to 'safe' limits, three broad activities must be pursued. These are efficient use of power, decarbonisation of electricity generation and the electrification of all transport.

10.3.1 Energy efficiency

Energy efficiency has been the low-hanging fruit of 'greening' strategies for decades when it comes to reducing energy demand. It will likely to remain so for several years yet. Studies on this topic [14,15] have generally concluded, bearing in mind the profligacy engendered by cheap fossil fuels (see Chapter 1), that energy efficiency improvements could net up to 50% of the emissions reductions required to meet global goals.

The demand for energy in human societies falls into four main categories. In 2050, on a BAU trajectory, industrial, commercial, and domestic consumption are predicted to absorb 75% of the total power generated, while transport will employ the remaining 25% [1]. Much of the power consumption in the industrial, commercial and domestic sectors is simply wasted because of poor building regulations and lax efficiency standards as suggested in Figure 10.3. For example [16] it has been demonstrated that, of the energy supplied to electricity generation stations in the form of fossil fuel or other energy sources, only 10% of it is actually used productively by the consumers. The rest dissipates as Joule heating in generators, transformers, transmission lines, and in user equipment and appliances. These inefficiencies in the 'electrical sector' of the economy are paralleled in all other areas where energy is expended for heating, lighting, powering office equipment and driving industrial machines. But it is the

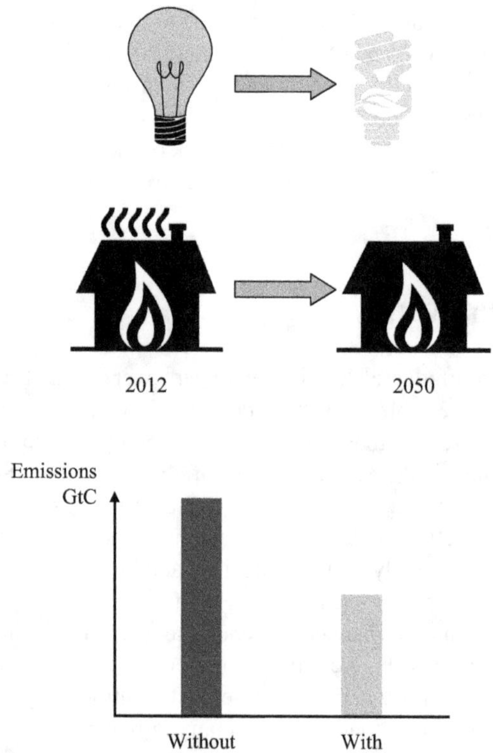

Figure 10.3 Implementing efficiency strategies showing relative emissions in 2050 'without' and 'with' aggressive efficiency drives

electrical savings that are particularly important since the future is heading towards 'all electric'. Clearly considerable savings are distinctly possible, but it will require concerted and sustained action to improve efficiency in all areas of power usage, such as heating, lighting, manufacturing equipment, farming equipment, power tools, electrical appliances, gas appliances, computers, office equipment, electronic devices etc., to realise the substantial savings which are possible.

It is surprisingly difficult to elicit relevant and helpful statistics from the literature, in order to form an accurate estimate of the energy savings, which could be made in the non-transport sectors. Perhaps, some clue as to what is possible can be surmised be comparing Switzerland and the USA, two industrialised nations with apparently similar gross domestic products (GDPs). On a per capita basis, Switzerland has been shown to use only 20% of the energy expended by the USA [17], to achieve a similar standard of living. Admittedly, this was several years ago – the difference may be less marked today. Furthermore, some studies suggest that the domestic sector in many parts of the world could be 80% more efficient than it is now [18,19]. Globally, therefore, it is not unrealistic to anticipate that coordinated and strenuous efforts at efficiency improvements, year-on-year, could reduce energy consumption by at least 40% by mid-century. This will require that the consumption of power in the non-transport sectors, currently supplied from fossil fuel burning plants, should decline by about 8%/year, simply by enforcing stringent efficiency standards and by making fossil fuel combustion very expensive. Ideally, total power consumption will remain flat at the 2012 level as renewables come 'on stream'.

10.3.2 Decarbonised electricity generation

Electrical power generation in industrial economies traditionally occupies about 12–15% of all energy use. However, as has already been indicated, electricity generation is not efficient, so we can expect that the savings in global carbon emissions by decarbonising the industry will be considerably more than this. Some studies have suggested up to 30%. In Figure 10.4, we conservatively suggest that a saving of 20% should be available by 2050 by a determined and extensive programme of conversion to 'clean' generation technologies around the globe.

The switch to renewable power for all our energy needs will have to incorporate a wide range of technologies. These will include hydro-electric power, wind power,

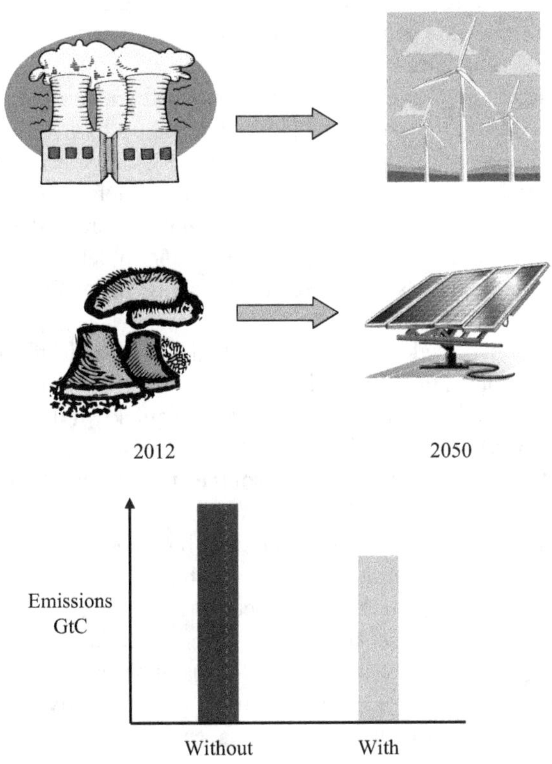

Figure 10.4 Emissions reductions derived from the decarbonisation of electricity supply

solar power, wave power, tidal power and geothermal power. Although not strictly 'renewable' since uranium is a finite resource, nuclear power is usually included in the mix, because it provides, at least in the short term, very useful levels of reliable base load. The transition will need to be coordinated on a continental scale, and will probably employ superconducting DC grid connections where large distances are involved. It is interesting to note that a 600 m long connection recently installed and demonstrated by the Long Island Power Authority in the USA was cooled by forcing liquid nitrogen through an annular space in the coaxial cable. The voltage was 138 kV and 2,000 A flowed in the cable. This represents the same power carrying capacity of a conventional 345 kV overhead line, but with virtually no transmission loss. In the future, modern materials offering room temperature superconductivity may dispense with the cooling that is not easy to sustain over long distances.

A feature of renewable power sources such as wind, wave and solar, which is raised repeatedly in debates about their capacity to replace fossil fuel-powered electricity generators, is intermittency of supply. However, at the global, or continental level (Europe say), the variability of renewables can more easily be addressed. When the wind is not blowing in Scotland, for example, it will likely be blowing in Germany! Under the auspices of the European Community, several reports have been generated to assess the feasibility of a direct current (DC) super-grid connecting geothermal power stations in central Europe, solar power stations in southern Europe and North Africa, wind farms in Western Europe, wave/tidal systems in Scandinavia and Portugal, and hydro-electric stations in Northern Europe. This system, in addition to base-load nuclear power, would be backed up by massive storage facilities based on compressed gas and hot water thermal storage using cathedral-sized underground caverns, on massive flywheel farms, on battery storage barns the size of football pitches and on huge super-cooled magnetic storage devices. Prototype examples of all of these technologies already exist, and undersea power lines across the Mediterranean Sea from North Africa to the continent of Europe are seriously being evaluated.

It should be emphasised that almost all of the renewable technologies listed above are relatively conventional. In principle, therefore, sustainable power systems based on these technologies could be introduced very quickly if drive, determination, enthusiasm and cooperation can be imbued in the international community, to recruit and deploy significant human and capital resources towards implementing the task. But where, at 'short notice', would the scientists and engineers required to implement the paradigm shift to renewables come from, and how could the required unprecedented expansion of manufacturing capability be achieved? The major components of renewable power stations, such as turbines, gear trains, generators, propeller blades, nacelles, control electronics, management systems, metering, mirrors etc., are, in engineering terms, not unlike what is currently manufactured in considerable volume by the automobile and aeronautic industries. Consequently, the answer to the above question is not too difficult to find if we accept that the future is largely *sans* oil. We must shift the manufacturing emphasis of these major factories, away from building, soon-to-be-redundant vehicles and aircraft, towards providing the infrastructure for renewable power plants, and we must use the capabilities of other fossil fuel dependent industries, such as those involved in chemicals and plastics, to develop storage systems and materials for a superconducting grid and we must aim to incorporate levitation technology into turbine bearings to make power systems as durable and maintenance-free as possible.

10.3.3 Electrification of transport

The largest share of greenhouse gas reductions from electrification will come from its application to transportation. Studies have shown that a fifth of all emissions could be saved by 2050 by a concerted drive to decarbonise transport. Today greenhouse gas emissions associated with the transport sector [21] are at about 15% of the total, and they are rising rapidly. On current trends, transport will contribute about 3 billion metric tons equivalent (3×10^{15} GtC/year) of greenhouse gases by 2050. This could fall to less than 0.2 billion metric tons with a determined and successful flight from fossil fuels. The change shown in Figure 10.5 represents a drop in emissions of about 20%.

Despite the loss of the 'products' of the automobile and aeronautic industries, *mankind* will not be reduced to *manpower* to get about. A power budget for transport could amount to not much less than it is now (in 2012) except that it will largely be supplied in the form of electricity. It will certainly allow a significant level of power-assisted travel, mainly in the form of ground-based mass people-movers, that is trains, ships, trams and trolley buses. Without treading too much into the area of future prediction, which is not a skill usually possessed by engineers, it seems appropriate here, to try to make some extrapolations, based on well-established technological trends. Hopefully by doing so, we can gain some understanding of what the major developments in the transport sector might be, when all energy comes in electrical form from renewables, and when, more importantly, it is severely capped at the 2012 level.

The biggest development, it is reasonably safe to say, will be in electrifying the railways. There will be much more of them serving a much wider community. The expansion of the railway system, perhaps including maglev, will become a high priority for governments once flying becomes no longer affordable, particularly high-speed international, and transcontinental, airways. The rapid expansion of rail systems may well take advantage of the emptying and freeing up of motorways as road traffic dwindles owing to the high cost or unavailability of fuel. Converting motorways to high-speed railways will be much easier than developing new networks. While this form of travel will be the primary replacement for air travel (Figure 10.5), for those that have to journey long distances relatively quickly, it is also easy to see that much of the need for roaming around the globe, that has been considered necessary in the past, is already being undermined by the massively improving accessibility of wideband communication systems and the Internet, through the agency of high-speed digital electronics. For example, electronic conferencing for large groups of people scattered around the globe will become commonplace, eliminating one incentive to travel for large numbers of individuals. In 30–40 years, electronic communications systems will be much more sophisticated than they are

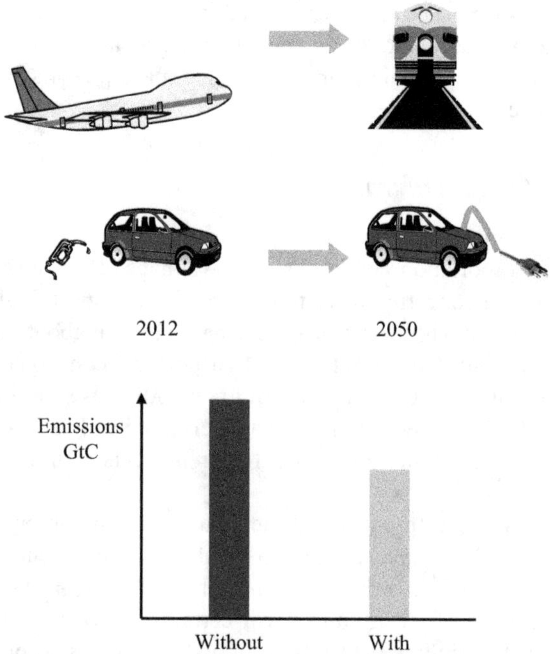

Figure 10.5 Emission savings associated with aggressive decarbonisation of transport

now, with computer processors continuing to increase in speed and memory capacity and broadband high-speed interconnections getting faster and more reliable. Clearly, many of the reasons for travel that existed in the past are being eroded.

Electrically powered commercial vehicles for delivery and other trading activities already exist within some cities and this trend to electrification will continue probably including vehicles such as taxis. Some may include solar panels to boost range. However, car ownership could possibly become a 'thing of the past' with electric town cars being available to hire on demand to those seeking journey flexibility and independence. Such a scheme (the Autolib system) has recently been pioneered in Paris [22]. Pedal power for local travel will very likely be the big beneficiary in towns and cities around the world. Electrically boosted bicycles will make this more possible in the hillier venues.

10.3.4 Societal savings

The above globally directed steps towards decarbonisation by massive efficiency drives and electrification are summarised in Figure 10.6 in the manner of Socolow and Pascala. The simplified linearised graphs representing emissions reductions from the BAU trend (diamonds) form three wedges representing efficiency (formed by area between BAU curve and triangles), decarbonising electricity supply (area between

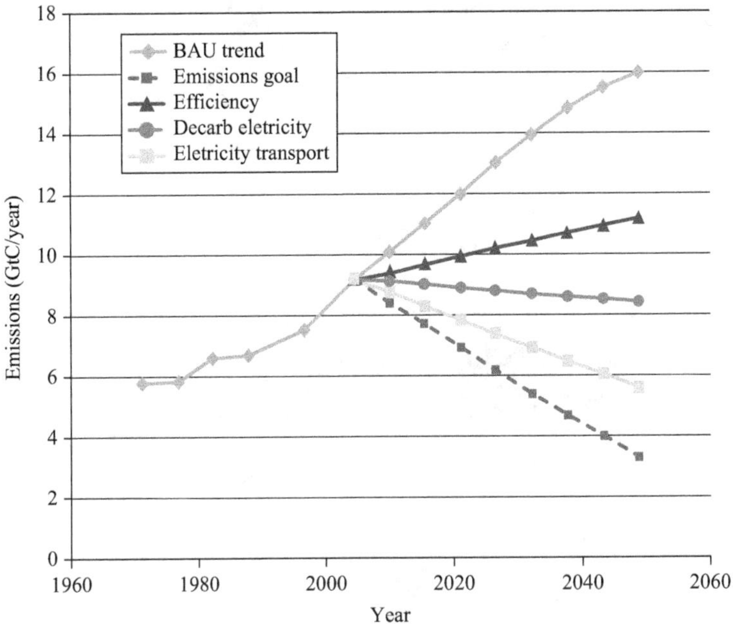

Figure 10.6 Emissions reductions associated with realistically exploitable strategies

triangles and circles) and electrifying transport (area between circles and squares). The remaining wedge (area between squares and the dashed curve) has to be filled by one or more additional slices in order to procure the missing 20% of further emissions reductions required to meet the climate change amelioration goal set for 2050.

Further real and substantial energy savings are possible, and many of these could probably be enforced by introducing a marginal energy pricing system, in which base-load electricity and gas/oil for essential requirements would be easily affordable, whereas for consumption demands beyond the base level the price/Joule would rise very steeply, so attracting increasingly expensive bills. How to do this at a global level is outside my area of expertise, and others with appropriate knowledge and skills will be required to devise a workable procedure. However, hopefully we would see disappear many uses of energy, especially in modern industrialised societies, which are frankly trivial and unnecessary. There are lots of examples, in the home, in entertainment venues, in the gymnasium, in the garden, in the workplace and elsewhere (Figure 10.7). At the time of writing, on one of the few days this summer, in the South East of Scotland, when the rain has stayed away and the sun has made a welcome appearance, the pleasure of decamping to the garden has been spoilt by noise pollution. The culprits are, of course, lawn mowers (mainly electric but petrol-driven version are also a pest), but today there is also an electric hedge trimmer grinding in the background. For able-bodied human beings why are such devices necessary?

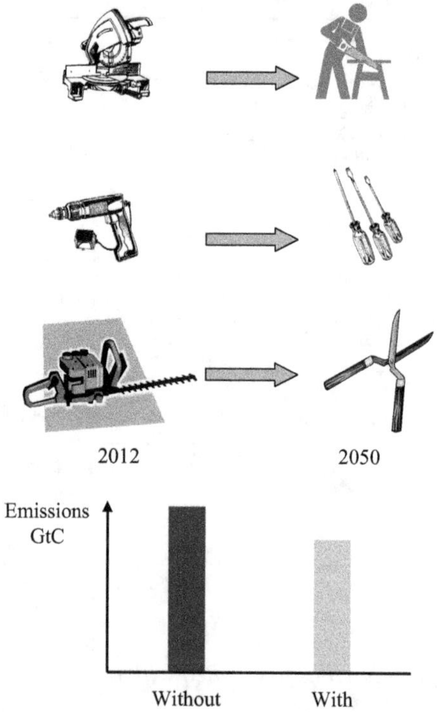

Figure 10.7 Emission reductions by lifestyle choices

Much less noisy push-mowers, and hand-operated hedge shears, were more than adequate to maintain the trim appearance of out-of-doors suburbia in the not too distant past. The manual versions also provided superb exercise for the user – surely a consideration in these days of spreading obesity?

Given that on average, during a working day, an adult human being is capable of providing muscle power of the order of 250 W [20], it is salutary to note that beyond 2050, a conservative 3 billion or so adult, able-bodied, men and women (~30% of the total population) on the planet, will represent available power for doing mechanical work of 0.75 TW. If all of this muscle power could be used to do work, which is currently being done by hand tools and other machines designed to boost human indolence, all powered by electricity, and remembering that electricity generation and transmission is, at best, 50% efficient, 1.5 TW (~10%) of renewable power generation could immediately be saved. This is the output of about 1,500 large power stations! With so much muscle power at our disposal, why do so many trucks, delivery lorries, removal vans, garbage collection vehicles seem to have hoists or cranes using the power of the engine to lift goods on and off the said vehicle, rather than use man power? The answer, of course, is easy access to ridiculously cheap fossil fuel energy. But in a resources strapped world, an awful lot of scarce energy can be saved by reintroducing muscle power. It was not so long ago, certainly with the memory span of anyone over 50, that coal delivery men were nonchalantly shifting 1 cwt coal bags on and off of trucks using their own 'brute strength'. The construction industry has also got rid of the 'muscle power' and the manual techniques that were more than good enough, in the not too distant past, to create the sophisticated buildings and structures appropriate to the needs of societies that were well advanced even by today's standards. Others are quite free to contemplate the further savings that could be procured by making intelligent and imaginative use of the muscle power of horses, elephants, yaks or oxen! Of course health and safety, and animal rights, issues would have to be addressed, but the rules may possibly change when energy is in increasingly short supply. It is, perhaps, pertinent to emphasise, that we are contemplating here the restoration of the health and safety of the planet itself, so it seems inevitable that unpalatable choices will have to be made at some stage!

Less controversially, savings can undoubtedly be procured by introducing clockwork, solar cell and perhaps kinetic mechanisms, into toys and electrical and electronic devices. Many free-standing electronic devices are increasingly being supplied with solar panels to power the electronics – such as calculators and watches. This could be extended to a much wider range of electrical components, as solar cells become more efficient and more robust. Apparently, a 40 W solar panel has recently been fitted to a hopefully quiet lawn mower [23], a clear indication that this technology has reached a stage where it is justifiable to suggest that significant savings in electricity usage globally, could soon be procured without seriously encroaching on individual liberties. My guess is that a further 10–15% saving in energy usage could be achieved, post 2050, by suppressing the worldwide manufacture and use of frivolous, mainly electrical gadgets, but also other unessential powered products. The object must be to increasingly introduce manual, solar-powered and clockwork powered devices and appliances into the market. Even higher savings of the order of 25% have been predicted for such programmes in a recent report from the McKinsey Global Institute [15].

It is likely that a final wedge could be implemented by changes in agricultural practices [13]. It has been found that up to 15% of the required emissions reductions could come from measures to reduce the not so insubstantial volumes of greenhouse gases that emanate from landfill and agricultural activities. Other possible sources of further emissions reductions are biofuels for the trucking industry and activities not easily served by electricity, smart planning of urban areas to minimise journey lengths and times for transport and rooftop solar photovoltaics.

10.4 Summary

The thrust of this chapter has been to show that, from an engineering perspective, the transition from an unsustainable global economy dependent on fossil fuels, to a sustainable one powered by renewables, is not impossible, and that technologies such as electromagnetic levitation and superconduction will form an important part of the drive to minimise energy wastage in a power constrained world.

We have noted that to effectively address climate change over the coming decade, the abandonment of fossil fuels will be required. This implies no less than a fundamental re-jigging of energy production, transportation and agriculture around the world. Since these are the sinews of modern life, a massive undertaking is in store – not unlike launching into a third world war. In addition to the flight from fossil fuels and the ensuing wholesale construction of renewable sources of energy, the pursuance of the 'war' will also involve the adoption of unprecedented levels of efficiency in engineering systems and devices ranging from power stations to wrist watches.

If the transition happens, and hopefully it will for the continued health of the planet, it will only do so if citizens around the globe demand it. How this can be socially engineered is for others to assess? Unfortunately, if mankind continues to follow the present BAU economic trajectory, the future for the biosphere looks decidedly bleak.

10.5 References

[1] *World Energy Consumption*. Available from www.solcomhouse.com/worldenergy.htm

[2] Meadows, D., Randers, J., & Meadows, D. *Limits to growth*. London: Earthscan Ltd.; 2005
If you are concerned about the planet and the future of mankind, this is essential reading. By marshalling a huge volume of real and hard data, and feeding it into a powerful software model of the global economy, scenarios for both economic collapse and sustainability are presented.

[3] Monbiot, G. *Heat*. London: Allen Lane; 2006
Although many of the statistics are geared to a UK readership, the science of global warming and the threat it poses to mankind, are well rehearsed in this trenchant contribution to the climate change debate.

[4] Homer-Dixon, T. *The upside of down*. London: Souvenir Press; 2006
The history of collapse of previous civilisations and empires, particularly the Roman Empire, which crumpled due to increasing shortages in energy in the form of food to feed the empire builders, is employed to predict the failure and disintegration of the American Empire, or more correctly the Western Industrial Empire. An analogy is made between the Roman food energy crisis and present day fossil fuel energy shortages. The Roman response is recruited in order to anticipate a likely growth in societal complexity to accommodate the changing circumstances and the fragility to external 'shocks' that this brings. It is postulated that knowledge of these trends may help us to avoid the total collapse of modern civilisation.

[5] Romm, J. *Hell and high water*. New York, NY: Harper Collins Publishers; 2007
With a strong background in science and energy politics in the USA, the author brings to the climate change debate a good combination of attributes. He ably addresses the challenges of transitioning to clean energy and escaping from

our addiction to fossil fuels. The hazards to life on Earth of not doing so are also elucidated cogently and powerfully.

[6] Flannery, T. *The weather makers*. London: Penguin Books; 2007
This is a clear and very readable exposition of the history, the problems and the consequences of climate change. The book is comprehensive encompassing as it does, possible causes of global warming , the growing evidence, and possible solutions, both of the techno-fix variety and those requiring societal changes. The transport dilemma arising from the lack of an alternative to fossil fuels is addressed, as is the urgent need to procure sustainable electric power.

[7] Lynas, M. *Six degrees*. Washington DC: National Geographic Society; 2008
'Six Degrees' paints a painful degree by degree picture of the changing planetary environment as the average global temperature, per degree centigrade, rises. Relying on the evidence from thousands of computer simulations, the deterioration of the living conditions, on land and in the sea, over most parts of the planet as the temperature rises inexorably, is skilfully described.

[8] Hansen, J. *Storms of my grandchildren*. London: Bloomsbury Publishing plc; 2009
Hansen spells out, very starkly and clearly, the severe dangers of ignoring scientific warnings on climate change, with the author's many years as a paleo-climatologist to give weight and emphasis to his arguments. His book is well written and refreshingly accessible to the non-scientist.

[9] Hamilton, C. *Requiem for a species*. London: Earthscan Ltd.; 2010
This book examines the choice currently facing mankind as to whether or not we act to save the planet for our grandchildren by striving to terminate those activities which are leading to devastating climate change. It acknowledges that 'we know the science' but that this understanding is largely undermined by much more powerful forces of denial, orchestrated by corporate power and money aided and abetted by bureaucratic inertia. The author's depressing conclusion is "Humanity failed".

[10] Sangster, A.J. *Warming to ecocide*. London: Springer-Verlag Ltd.; 2011
Here the science of global warming is approached from an engineering perspective with non-equilibrium, thermodynamics as the primary tool. Given that thermodynamics has been well understood by scientists and engineers for almost 200 years it is suggested that mankind has been seriously remiss in failing to address the climate change problem many decades ago.

[11] Gilding, P. *The great disruption*. London: Bloomsbury Publishing plc; 2011
The thrust of this response to climate change is a contemplation of the loss, suffering and disruption that will be inflicted on humanity as it attempts to come to terms with a warming planet. Nevertheless, the book optimistically suggests that by exercising resilience and adaptability, in reshaping the global economy away from consumerism and growth, it is still possible for mankind to avert a total catastrophe.

[12] International Panel on Climate Change (IPCC). *4th Assessment Report*; 2007

[13] Pacala, S., & Socolow, R. 'Stabilization wedges: solving the climate problem for the next 50 years with current technologies'. *Science*. 2004;**305**:968–971

[14] *World Energy Consumption*. Available from www.solcomhouse.com/worldenergy.htm

[15] Bressand, F., Farrell, D., Haas, P., Morin, F., & Nyquist, S. *Curbing global energy demand growth: the energy productivity opportunity*. San Franscisco: McKinsey Global Institute, May Report, 2007

[16] Sangster, A.J. *Energy for a warming world*. London: Springer-Verlag Ltd.; 2010

[17] Brockris, J.O'M. *Energy options*. London: Taylor Francis Ltd.; 1980

[18] Tickell, O. *Kyoto 2: how to manage the global greenhouse*. London: Zen Books; 2008

[19] *Breaking the climate deadlock*. G8 Report, Japan, 2008

[20] Available from http://en.wikipedia.org/wiki/Human_powered_vehicle

[21] Synthesis report of the IPCC. *4th Assessment Report, IPCC*, November 2007

[22] Available from www.bbc.co.uk/news/world-europe-15134136

[23] Available from http://www.sustainableliving.com.au/profiles/stories/solar-powered-lawn-mower/

Appendix I

Lumped electrical circuit elements

While Maxwell's equations are applicable to virtually all macroscopic electrical phenomena, they are often mathematically too difficult to apply directly to realistic structures. Circuit board elements, such as apparently simple capacitors and inductors, come into this category. However, for such elements, provided they can be considered to be very small relative to the wavelength in free-space at the operating frequency of the electrical system, an analysis technique termed lumped element circuit theory furnishes powerful insight into the behaviour of quite complex multi-component networks.

Capacitance

The relationships between fields, forces and energy storage in electrostatic systems can best be constructed by application of the first Maxwell equation (2.13) to a simple parallel plate capacitor, as depicted in Figure A1.1. It comprises a pair of square metal electrodes (area A), in a parallel arrangement, and separated by a distance t. The plates are held apart by an insulator of relative permittivity ε_r.

For a simple parallel plate capacitor for which t is very small by comparison with the electrode dimensions, it is possible to assume with little error, that the superimposed electric field exists wholly between the plates, with negligible fringing

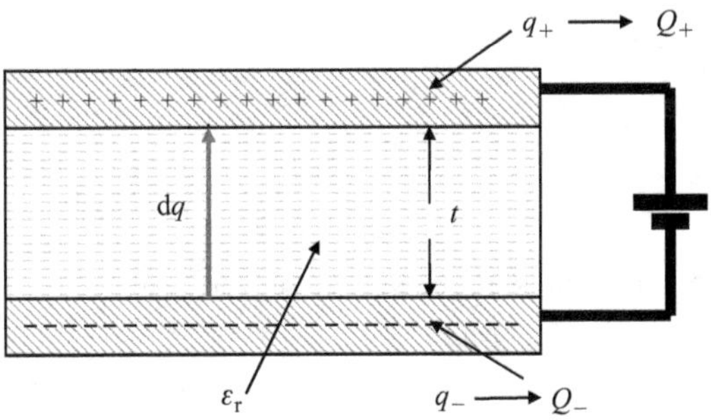

Figure A1.1 Parallel plate capacitor charged to voltage V resulting in charge Q_+ in top plate and Q_- on the lower plate

into the exterior region. Furthermore, this field **D** is uniformly distributed and constant across the gap. In this case, the applicable Maxwell equation (2.13) when applied to the box-shaped surface (Gaussian surface) enclosing the top plate leads to

$$DA = Q_+ \tag{A1.1}$$

But the constitutive relation (Chapter 2) tells us that $D = \varepsilon_r \varepsilon_0 E = \varepsilon E$. Hence

$$E = \frac{Q_+}{\varepsilon A} \tag{A1.2}$$

Now given that voltage is defined as the line integral of **E** between any two points in the field (Chapter 2, reference 1), we must have $V = Et$ (in Volts). Consequently

$$V = \frac{Q_+ t}{\varepsilon A} \quad \text{(Volts)} \tag{A1.3}$$

So, it is clear that the ratio

$$\frac{Q}{V} = \text{constant} \tag{A1.4}$$

This constant is defined as the capacitance C (in Farads) of the system (Chapter 2, reference 1). Hence we must have, for the parallel plate capacitor

$$C = \frac{\varepsilon A}{t} \quad \text{(Farads)} \tag{A1.5}$$

Equation (A1.4) implies that

$$V = \frac{Q}{C}$$

Hence on differentiation both sides of the equation with respect to time we obtain

$$\frac{\mathrm{d}V}{\mathrm{d}t} = \frac{1}{C}\frac{\mathrm{d}Q}{\mathrm{d}t} = \frac{i}{C} \tag{A1.6}$$

where i is the instantaneous current flowing into the capacitor. If the source V is a generator rather than a battery operating at frequency ω (Hz), so that $V = V_0 \exp(j\omega t)$, the above equation can be recast in circuit form as

$$V = \frac{i}{j\omega C} \quad \text{(Volts)} \tag{A1.7}$$

Also

$$X_C = \frac{1}{j\omega C} \tag{A1.8}$$

is termed the reactance in ohms of the capacitor circuit element.

Inductance

In Figure A1.2, the N-turn coil wound onto a toroidal iron core of permeability μ_r is connected to a high-impedance current source. The high impedance ensures that any back e.m.f. in the coil has little influence on the instantaneous generator current i.

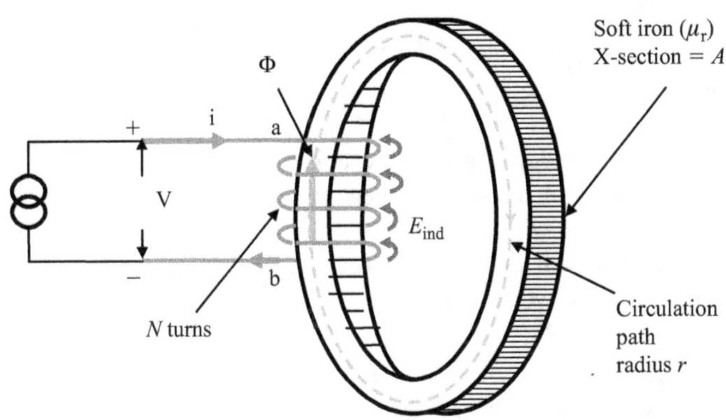

Figure A1.2 Induced voltage in coil

If we now apply Ampere's circuital law (Chapter 2) around a circulation path within the iron core (see diagram), the following approximate relationship is deduced:

$$\text{circ}\,\mathbf{H} = Ni \tag{A1.9}$$

For a path at a median radius r within the core this yields

$$2\pi r H_\theta = Ni \tag{A1.10}$$

which implies that

$$H_\theta = \frac{Ni}{2\pi r} \tag{A1.11}$$

We are now in a position to compute the magnetic flux (Φ, webers) in the core, by employing the relations

$$B_\theta = \mu_r \mu_0 H_\theta \tag{A1.12}$$

and

$$\Phi = B_\theta A \tag{A1.13}$$

assuming the core exhibits a uniform cross-sectional area A (m^2).

Furthermore, the integral form of Faraday's law (Chapter 2) is applicable to a circulation path, which proceeds from point a to point b within the wire forming the coil, then from b to a in the air. For this path, we obtain

$$\text{circ}\,\mathbf{E} = -\frac{\mathrm{d}}{\mathrm{d}t}(N\Phi) \tag{A1.14}$$

Within the perfectly conducting wire, circ $\mathbf{E} = 0$, therefore the above equation becomes

$$\text{circ}\,\mathbf{E}|_{ba} = -N\frac{\mathrm{d}\Phi}{\mathrm{d}t}$$

or

$$V = N\frac{\mathrm{d}\Phi}{\mathrm{d}t} \tag{A1.15}$$

where V is the voltage induced across the coil by the time varying current i. The negative sign disappears since point b is negative with respect to point a, for increasing Φ in the direction shown.

From (A1.11) it is clear that $H_\theta \propto i$, consequently, for an iron core that is linear (μ_r independent of i) that we must have

$$N\Phi \propto i$$

When a constant of proportionality L is introduced we obtain

$$N\Phi = Li \tag{A1.16}$$

L is referred to as the inductance of the coil in Henries, and is given by

$$L = \frac{N\Phi}{i} \quad \text{(Henries)} \tag{A1.17}$$

The lumped circuit form of (A3.14) is obtained by combining (A1.15) and (A1.16) that results in the relationship

$$V = L\frac{\mathrm{d}i}{\mathrm{d}t} \quad \text{(Volts)} \tag{A1.18}$$

Again, if the current source generates a sinusoidally varying current with $I = i_0 \exp(j\omega t)$, then

$$V = j\omega Li \quad \text{(Volts)} \tag{A1.19}$$

and $X_L = j\omega L$ is the inductive reactance in ohms.

Resistance

Conductivity (σ, in Seimens), in real electrical materials, is a measure of the proclivity of the atomic structure of the metal, semiconductor or insulator to furnish, and permit the flow of, free electrons. It is large in metals and virtually zero in insulators. Conductivity appears in the Maxwell relations through (2.15) and in particular the I_{cond} term. In a long, straight resistive wire of length l and cross-sectional area A, at very low frequency or at DC, the current density in the wire J is given approximately by

$$J = \frac{I_{\text{cond}}}{A} \quad (\text{A/m}^2)$$

But we also have (see Chapter 2, reference 1)

$$J = \sigma E$$

where E is the electric field stress within the wire. Hence given that for a uniform wire the voltage between the ends of the wire (V) is (see Chapter 2, reference 1):

$$V = \int_0^{\ell} E\mathrm{d}\ell = E\ell = \frac{I_{\mathrm{cond}}\ell}{\sigma A} = I_{\mathrm{cond}}\frac{\rho\ell}{A} \quad \text{(Volts)} \tag{A1.20}$$

Here, $\rho(=1/\sigma)$ is the resistivity of the wire in ohms-m. Clearly, the above equation can be expressed in the form

$$V = I_{\mathrm{cond}}R \tag{A1.21}$$

where the resistance R (Ω) has the form

$$R = \frac{\rho\ell}{A} \tag{A1.22}$$

Equation (A1.21) is Ohm's circuital law.

Appendix II
Electrostatic force calculations

The calculation of the suspension force for the four-electrode electrostatic suspension system described in Chapter 5 is made easier by making an assumption commonly employed in capacitor problems of uniform electric field between an electrode and the float. In which case we can state that

$$C_i = \frac{\varepsilon_r \varepsilon_0 A_i}{g_i} \tag{A2.1}$$

where A_i is the overlapping area between the *i*th electrode and the displaced disk. It is clearly a function of the off-set x_s (see Chapter 5). If we define a displacement function f_k as follows:

$$f_k = +\sqrt{(R_k^2 - (x - x_s)^2)} \quad \text{with } k = 1, 2 \tag{A2.2}$$

Hence A_i has the following integral form:

$$\begin{aligned} A_i(x_s) = & \int_{\gamma}^{\lambda} f_1(x, x_s)\mathrm{d}x + (-1)^i \int_{x_s}^{0} f_1(x, x_s)\mathrm{d}x \\ & - \left\{ \int_{\lambda}^{\alpha} f_2(x, x_s)\mathrm{d}x + (-1)^{i+1} \int_{\delta}^{x_s} f_2(x, x_s)\mathrm{d}x \right\} \end{aligned} \tag{A2.3}$$

where it is assumed that $-R_2 < x_s < R_2$ and

$$\gamma = -R_1 + |x_s|, \quad \lambda = \frac{|x_s|}{2}, \quad \alpha = R_2 + |x_s|, \quad \delta = \frac{x_s}{2}$$

The suspension force on the float in Figure 5.4 due to all four electrodes can be generated by combining (5.14) and (A2.1) to yield (Chapter 5, reference 6):

$$\begin{aligned} F_z &= \sum_{i=1}^{4} F_{z,i} \\ &= \varepsilon_r \varepsilon_0 \sum_{i=1}^{4} \frac{A_i}{g_i^2} \left\{ V_i - \frac{\sum_{j=1}^{4} \frac{A_j V_j}{g_j}}{\sum_{j=1}^{4} \frac{A_j}{g_j}} \right\} \\ &\quad \times \left\{ \frac{A_i}{g_i} \frac{\left[\sum_{k=1,k\neq i}^{4} \frac{V_k A_k}{g_k} - V_i \sum_{m=1,m\neq i}^{4} \frac{A_m}{g_m} \right]}{\sum_{k=1}^{4} \sum_{i=1}^{4} \frac{A_k A_i}{g_k g_i}} - \frac{1}{2} \left[V_i - \frac{\sum_{j=1}^{4} \frac{A_j V_j}{g_j}}{\sum_{j=1}^{4} \frac{A_j}{g_j}} \right] \right\} \end{aligned} \tag{A2.4}$$

Although superficially complicated, the repetitive nature of this equation makes it easy to translate into an embedded computer algorithm to provide rapid force calculations within a changing micro-electro-mechanical system model.

Symbols and abbreviations

A	ampere
AC	alternating current
A/D	analogue to digital
AMB	active magnetic bearing
BAU	business-as-usual
BTU	British thermal unit
c	speed of light (3×10^8 m/s)
c-	centi- ($\times 0.01$)
C	Coulomb
CEM	classical electromagnetism
CW	continuous wave
DC	direct Current
DSP	digital signal processor
ECL	Eddy current levitation
EM	electromagnetic
EML	electromagnetic levitation
ESL	electrostatic levitator
η	efficiency
emf	electromotive force
ε	permittivity (Farads/m)
F	Farad
g	gram
g	gravitational acceleration (9.807 m/s^2)
G-	giga ($\times 10^9$)
GDP	gross domestic product
GtC	gigatons carbon
H	Henry
HAWT	horizontal axis wind turbine
Hz	Hertz (= cycles/second)
HTS	high temperature superconductor
$\hbar$	Plank's constant
k-	kilo ($\times$1,000)
K	Kelvin
LC	inductor (L) plus capacitor (C)
LED	light emitting diode
λ	wavelength

m-	milli ($\times 0.001$)
m	metre
maglev	magnetic levitation
M-	mega ($\times 10^6$)
μ-	micro ($\times 10^{-6}$)
μ	permeability (Henry/m)
MEMS	microelectromechanical systems
m.k.s.	metre.kilogram.second
mmf	magnetomotive force
mph	miles per hour
n-	nano ($\times 10^{-9}$)
N	Newton
NASA	National Aeronautical Space Administration
ω	frequency (rad/s)
Ω	Ohm
p-	pico ($\times 10^{-12}$)
ppmv	parts per million by volume
PWM	pulse width modulation
Q	quality factor
QED	quantum electrodynamics
rev	revolution
r.m.s.	root mean square
s	second
σ	conductivity (mhos-m)
SLIM	single-sided linear induction motor
T-	tera ($\times 10^{12}$)
T	Tesla
TE	transverse electric
TEM	transverse electromagnetic
TM	transverse magnetic
UMP	unbalanced magnetic pull
UV	ultra-violet
V	Volt
VAWT	vertical axis wind turbine
W	Watt
yr	year

Index

www.ingramcontent.com/pod-product-compliance
Lightning Source LLC
LaVergne TN
LVHW011758200926
848484LV00003B/730
* 9 7 8 1 8 4 9 1 9 6 6 3 5 *